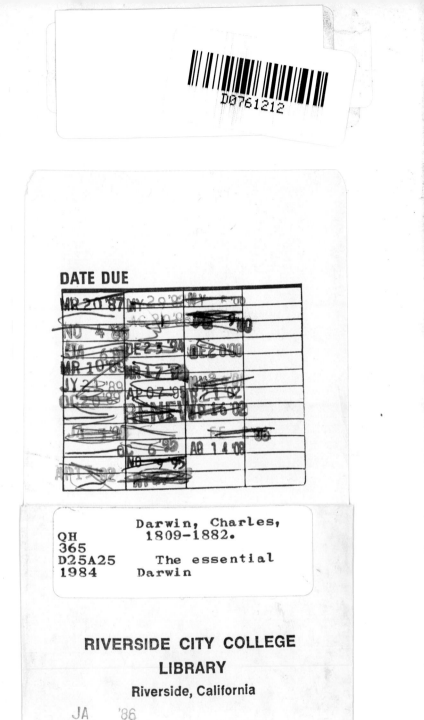

 *THE ESSENTIAL DARWIN*

MASTERS OF MODERN SCIENCE SERIES

Also by Robert Jastrow

RED GIANTS AND WHITE DWARFS

ASTRONOMY: FUNDAMENTALS AND FRONTIERS
(WITH MALCOLM THOMPSON)

UNTIL THE SUN DIES

GOD AND THE ASTRONOMERS

THE ENCHANTED LOOM

MASTERS OF MODERN SCIENCE SERIES

Robert Jastrow, *General Editor*

The Essential
DARWIN

Selections and Commentary by
Kenneth Korey

LITTLE, BROWN AND COMPANY   BOSTON   TORONTO

LICATION DATA

The essential Darwin.

(Masters of modern science series)
Bibliography: p.
Includes index.
1. Darwin, Charles, 1809–1882. 2. Evolution.
3. Natural history. 4. Naturalists — England — Biography.
I. Jastrow, Robert, 1925–   II. Korey, Kenneth A.
III. Title.   IV. Series.
QH365.D25J37   1984   575.01'62   84-15502
ISBN 0-316-45826-0
ISBN 0-316-45827-9 (pbk.)

MV

DESIGNED BY DEDE CUMMINGS

*Published simultaneously in Canada
by Little, Brown & Company (Canada) Limited*

PRINTED IN THE UNITED STATES OF AMERICA

Contents

Preface

THE CONTRIBUTION of Charles Darwin to the architecture of modern thought hardly needs introduction; no other single achievement in science has produced an impact on such seismic scale. Darwin delivered an idea whose power was to rearrange the conceptual world-order of Western society. The central theme of that order, Providentially supervised Creation, had endured the cosmological revisions of Copernicus and Galileo, but it could not withstand evolution by natural selection. Science after Darwin no longer calls upon miraculous intervention to explain the workings of the natural universe, nor — more momentously — does mankind still reside at its center. Indeed, even a century after his death, the aftershock of Darwin's science registers in intermittent rumblings from the theologically conservative.

An accomplishment so monumental as Darwin's has left, quite expectably, a great wake of interpretative scholarship. Then why, as a colleague recently asked, yet another book on Darwin? My answer is that this is much less a book about Darwin than it is Darwin's book. I am convinced that there remains great value in reading Darwin, just as I am equally convinced that too few do. Perhaps this is because the nub of Darwinism — evolution by natural selection — appears so marvelously transparent in its conception as to discourage further burrowing for its root. This is unfortunate, for the Darwin who inhabits biology textbooks — the vulgar Darwin — is a pallid representation of the Darwin who lived and whose writings are contained in these pages. True, both Darwins returned from the *Beagle* voyage having observed species so arrayed as to suggest descent with modification from common ancestors, and both conjoined Malthusian superfecundity under limited resources with inherited diversity, thus arriving at natural selection to power evolution. But did this vulgar Darwin also share his real-life counterpart's enduring and expanding commitment to Lamarckian inheritance, the inheritance of acquired characteristics? Almost never did he, and if ever, then only a "brief flirtation," an "occasional reference," or, worse, only an "initial" reliance because of his unfamiliarity with random sources of variation (thereby compounding two misstatements). Chameleonlike, the vulgar Darwin is at ease with the science of the present moment; he has adroitly skirted the obstacles of a science in its formative stage.

To be sure, there is much in Darwin's work that holds up gracefully to the passage of time, just as there is much long ago discarded. Darwin himself was poignantly aware of his theory's difficulties and its want of revision, for shortly before his retirement he confided to his close friend, Joseph Hooker,

> If I lived twenty more years and was able to work, how I should have to modify the *Origin*, and how much the views on all points will have to be modified! Well it is a beginning, and that is something. . . .[1]

I do not know exactly why Darwin came to be bowdlerized. Perhaps the requirement for heuristic simplicity is the only explanation needed, although my dreariest suspicion, founded upon the factual inaccuracies that often accompany the vulgar Darwin, is that many writers are limited in their acquaintance with the original. Whatever the case, the vulgar version of Darwinism is so firmly established in our folklore that to attend too closely to Darwin's mid-Victorian misconceptions lends an appearance of reckless sensationalism. I shall run that risk here, because I believe more is to be learned by examining first-hand the self-corrective dialectic of science than by glossing over those details that history discarded from its modern editions. Besides, the progress of Darwin's theory from its first appearance to its final revisions is intelligible only with reference to his varied deployment of these mistaken notions.

The foremost goal of this volume, then, is to introduce Darwin's writings to a larger audience. To accomplish this I have tried to allow Darwin to tell his own story as far as possible. I have interlaced his text with numerous overviews and critical notes, but the center stage is clearly his. Darwin's scholarly output was prodigious in its volume and breadth (doubtless his decision to pass by a university post in favor of a research career was in his case a wise one). His writings thus offer his editors great latitude in determining how best to convey the sense of his enterprise. Sacrificing diversity to gain coherence, I have elected to present an abridgment of his most significant work, *On the Origin of Species*, along with abridged selections from several other volumes. I have chosen the first edition (1859) of *Origin* to represent his theory as it was originally offered, before it underwent successive modifications over the course of more than a decade. To help illuminate its trajectory I have selected an important chapter from the second edition (1874) of *The Descent of Man*, Darwin's last major theoretical contribution.* I have let Darwin introduce himself through his *Autobiography*. Although too great a reliance upon self-appraisal is risky, his recollections evoke aptly the milieu of his formative experience. Finally, because of its prominence in Darwin's seminal vision of evolutionary change, I have included from the *Voyage of the Beagle* the chapter on the Galapagos Archipelago.

While presenting Darwin is the primary goal here, I have introduced a subordinate theme to serve as an integrative ma-

---

* In fact, the choice of the second edition of *Descent* is more to preserve symmetry than anything else, since its text differs unimportantly from the first edition's.

trix. I have added several brief chapters to chronicle the progression of Darwin's conception of evolution over the course of his career and then, thereafter, as it advanced to the present through the custody of others. Although the format of this volume limits full development of this theme, it is my hope to provide some sense of how science proceeds. As a voluminous scholarship indicates, there are many productive ways to examine Darwin's idea and its multiplicity of ramifications. His copious notes and diaries lend themselves well to biography, as de Beer[2] and Gruber[3] have so successfully shown. Or one might consider the wider social reverberations of Darwinism, as Himmelfarb[4] and Hofstadter[5] have done, or the radically new method Darwin sponsored for doing historical science, as Ghiselin[6] has done, and on and on. While the nature of this volume restricts the scope of such considerations, I have tried to direct the reader to further, more comprehensive treatments.

I have attempted in abridging Darwin's texts to employ as light a hand as possible. This has been a painful but necessary task, recognizing that recitative to me may be aria to another. I have first excised footnotes and numerous references to writings and debate of only ephemeral importance. Next, more reluctantly, I have cropped examples Darwin gave to support one or another argument when these seemed excessively redundant. On the other hand, I have often left intact, where the importance of the issue warrants, blocks of supporting evidence in order to preserve the form of argument Darwin employed. Where I have eliminated illustrative material I have tried to omit the less apt examples, and where all seemed equally suitable I exercised my personal predilections. I am an anthropologist and evolutionary biologist; a botanist would likely have made different choices.

The Background
to Darwin's Theory

DARWIN'S BOOKS belong to the category of great works that are much discussed but little read. Yet these works contain the most arresting and significant set of ideas relating to man and to his surroundings that has ever come from the scientific mind. It is our good fortune that Professor Kenneth Korey has been willing to undertake the task of pruning away the denser elements in Darwin's writing, leaving exposed the great structures of his thoughts on the nature of life for all to admire and understand.

Darwin has been more honored and more reviled than any other scientist since Galileo. What did he do to earn such praise and hostility? First, he proved the fact of evolution. Most people in his time believed, and some still believe today, that every kind of animal on the earth has appeared here through an act of special creation. Darwin said this was not so. He amassed convincing circumstantial evidence that all forms of life on the earth have evolved out of other forms that lived at an earlier time, and that those in turn had evolved out of still earlier forms. The chain of life extends far back into the distant past to some ancestral stand of matter lurking in a warm pool. Every creature on our planet is a cousin to every other, according to Darwin; all are related to one another by descent from common ancestors. Man is among these creatures; he, too, has evolved out of lower and simpler kinds of life and differs in no fundamental way from other animals on this planet. "Man," Darwin wrote, "is descended from a hairy, tailed quadruped, probably arboreal in its habits."

This last statement came as a particular shock to Darwin's contemporaries. As Copernicus had removed the Earth from the center of the Universe three centuries earlier, now Darwin removed man from the center of the world of living things. In that achievement he joined the ranks of great men — Newton and Einstein, among others — who have contributed to the modern understanding of man's place in the cosmic order.

Darwin's second great achievement was the discovery of the law that governs evolution. Granted the fact of evolution, how does it happen? And why does it produce one particular plant or animal, and not another? Why the whale? The elephant? The monkey? Why not some completely different sort of beast? Before Darwin, divine action was the only answer known to these questions. Darwin's genius revealed another explanation, one that required no supernatural intervention. He discovered a new law in nature that molds the forms of life as firmly as the force of gravity molds the stars and planets. *

According to Darwin's law, evolution works on plants and animals through differences in their "reproductive success" — i.e., the number of offspring each individual produces. He reached this conclusion on the basis of three facts about living things, and a straightforward line of reasoning.

The first fact is *exponential growth* — the tendency of all living things to increase rapidly in numbers. Darwin was enormously impressed by this property of life. He wrote in the *Origin of Species,*

---

* It seems puzzling at first that a new law had to be discovered. Since all living things are made of simple atoms that interact according to well-worked-out laws of physics, it would seem that the methods of the physicist, developed in enough detail, should be able to account for every variety of plant and animal on the earth. It turns out, however, that even the most primitive forms of life are too complex to be analyzed in this way. The physicist's methods, which have yielded such a deep understanding of inanimate matter, are powerless to explain the subtleties of biological evolution.

There is no exception to the rule that every organic being naturally increases at so high a rate, that, if not destroyed, the earth would soon be covered by the progeny of a single pair. Even slow-breeding man has doubled in 25 years, and at this rate, in less than a thousand years, there would literally not be standing room for his progeny.

The second fact is *variation* — all elephants look more or less alike; human beings look like one another; but within the population of any particular kind of plant or animal, small variations exist from one individual to another.

The third fact used by Darwin is *inheritance*. It is a property of all living things that offspring tend to inherit the traits of their parents; for example, tall parents tend to have taller-than-average offspring and fair parents tend to have fair-skinned offspring.

Building on these three observations about the nature of life, Darwin constructed his theory of evolution by a tightly reasoned argument: (1) a population grows until it approaches the limit of its resources; (2) in the struggle for existence that results, individuals with traits that help them to overcome the adverse forces of the environment — famine, disease, a harsh climate and the attacks of the predator — are more likely to survive and produce offspring; (3) the offspring tend to inherit the favorable traits from their parents, and carry them on into future generations; (4) on the other hand, individuals with traits that handicap them in the struggle against adverse forces are less likely to reach maturity and therefore less likely to produce offspring, and their traits will tend to disappear from the population; (5) over the course of many generations, this process, which preserves and strengthens some traits while it prunes away others, gradually transforms the species.

To the eye of the observer who only sees in the fossil record the beginning and the end of this particular segment in the history of life, and not the imperceptibly slight modifications from one generation to the next that intervened, it appears that a new species has suddenly been created. Darwin wrote in the *Origin of Species:*

> The mind cannot grasp the full meaning of the term of even a million years; it cannot add up and perceive the full effects of many slight variations, accumulated during an almost infinite number of generations. . . . We see nothing of these slow changes in progress, until the hand of time has marked the lapse of ages, and then . . . we see only that the forms of life are now different from what they formerly were.

Darwin gave to this process — the creative power that generates new forms of life — the name *natural selection.* He wrote in the *Origin of Species,*

> Natural selection is daily and hourly scrutinising, throughout the world, the slightest variations; rejecting those that are bad, preserving and adding up all that are good; silently and insensibly working . . . at the improvement of each organic being in relation to its organic and inorganic conditions of life.

There it is in a nutshell, the mechanism of evolution: the principle that accounts for the trunk of the elephant, the neck of the giraffe and the brain of man, inferred from observations on the nature of life that could have been made by anyone. How did Darwin arrive at this result, so powerful and yet so easily grasped? Evolutionist Ernest Mayr suggests that the ingredients of Darwin's greatness were

> A brilliant mind, great intellectual boldness, and an ability to combine the best qualities of a naturalist-observer, of a philosophical theoretician, and of an experimentalist. The world has so far seen such a combination only once and this accounts for Darwin's unique greatness.

According to Darwin, he found the catalyzing element for his theory in an essay by Malthus, a Scottish clergyman who stressed the tendency of populations to increase rapidly in numbers without limit, and the harsh struggle for survival that resulted. In Darwin's autobiography he wrote:

> Fifteen months after I had begun my systematic enquiry (that is, in 1838), I happened to read for amusement Malthus on Population, and being well prepared to appreciate the struggle for existence which everywhere goes on, from long-continued observation of the habits of animals and plants, it at once struck me that under these circumstances favorable variations would tend to be preserved, and unfavorable ones to be destroyed. The results of this would be the formation of new species. Here, then, I had at last got a theory by which to work.

Thus was the germ of the great idea planted in Darwin's mind. Twenty-one years later, the fully developed theory appeared in the *Origin of Species*. Its essentials are elegantly summarized in the following passage from the *Origin*:

> A struggle for existence inevitably follows from the high rate at which all organic beings tend to increase. Every being . . . must suffer destruction during some period of its life. . . . Otherwise, on the principal of geometrical increase, its numbers would quickly become so inordinately great that no country could support the product. Hence, as more individuals are produced than can possibly survive, there must in every case be a struggle for existence, either one individual with another of the same species, or with the individuals of distinct species, or with the physical conditions of life. It is the doctrine of Malthus applied with manifold force to the whole animal and vegetable kingdoms. . . .

> Owing to this struggle [for existence], variations, however slight . . . if they be in any degree profitable to the individuals of a species . . . will tend to the preservation of such individuals,

and will generally be inherited by the offspring. The offspring, also, will thus have a better chance of surviving, for, of the many individuals of any species which are periodically born, but a small number can survive. I have called this principle, by which each slight variation, if useful, is preserved, by the term Natural Selection, in order to make its relation to man's power of selection. But the expression often used by Mr. Herbert Spencer of the Survival of the Fittest is more accurate and sometimes equally convenient.

TWO MYTHS ABOUT DARWIN'S THEORY

No scientific idea, except perhaps the theory of relativity, has been the source of as much misunderstanding as natural selection. Much of the confusion relating to Darwin's ideas originates in the graphic phase, Survival of the Fittest. This was, in fact, as Darwin noted, not Darwin's invention, but that of his contemporary, the philosopher Herbert Spencer. Evidently, Darwin felt that Spencer's phrase captured the essence of his theory, because he borrowed it from Spencer and inserted it into the *Origin of Species* at several points. From that choice followed much mischief, for when the question is asked, "In the phrase, Survival of the Fittest, who are the fittest?" The answer comes back, "Those who survive." Thus, the central concept in natural selection is revealed to be "the survival of those who survive," and the theory is reduced to a meaningless tautology.

Or so it would seem. But this reading of Darwin is based on a misunderstanding of his ideas. In the theory of natural selection, fitness has a very special meaning: a fit individual is not merely one who survives, but one who also produces offspring. Darwinian fitness means *reproductive success*. A person may have great physical strength, nobility of character, and brilliance of intellect, but if he or she produces no offspring, that person's fitness is zero.

The distinction is made clear by the example of the praying mantis. After the male praying mantis has fertilized the female's eggs, she begins to eat him, starting with his head. Sometimes he gets away, but more often not. Moreover, the brain of the male secretes a hormone that inhibits copulation; but after the male has lost his head, he copulates more vigorously. These circumstances enhance the male's reproductive success, contributing greatly to the survival of the species, but severely diminish his personal prospects of survival. Although this example is drawn from the insect world, similar considerations apply to the higher animals and man.

Another source of confusion in thinking about evolution relates to the role of chance. According to the current scientific view, chance collisions between molecules led to the appearance of the first forms of life; then chance variations in the forms of life led to the evolution of complex forms out of simple ones; finally, after four billion years, man appeared, emerged out of a succession of more primitive forms, and seemingly the product of a long chain of random events. Can this be true?

For most people, such an explanation of the origin of mankind is as improbable as the creation of the *Mona Lisa* by the random splatter of paints on a canvas. But few scientists share this skepticism. The majority of scientists feel that Darwin's law of natural selection removes the need for a guiding hand in the Universe; in their view the theory of evolution is complete and requires the action of no mysterious forces beyond the reach of scientific understanding. The great evolutionist George Gaylord Simpson wrote that evolution "achieves the aspect of purpose without . . . a purposer, and has produced a vast plan without . . . a planner."

Yet, when you step back to look at the history of life with the perspective of hundreds of millions of years, you see that it has a flow and a direction — from simple to complex, from lower forms to higher — and you wonder: Can this history of events leading to man, with its clear direction, yet be undirected? Darwin himself was uncertain about the matter. In his autobiography he noted that when writing the *Origin of Species*, he was overwhelmed by "the extreme difficulty or rather impossibility of conceiving this immense and wonderful universe . . . as the result of blind chance," and when thus reflecting, he went on, "I feel compelled to look to a First Cause." Later his belief in the need for postulating a creative force weakened, and in 1870 he confessed: "My theology is a simple muddle. I cannot look at the universe as the result of blind chance, yet I see no evidence of beneficient design in the details." Finally, toward the end of his life, he retreated to the middle ground, and wrote: "I for one must be content to remain an agnostic."

In my opinion, Darwin's final agnosticism is the only appropriate reaction a scientist may have to these questions of teleology in science. Scientists can neither prove nor disprove that man is the product of chance; the cause *of* a chain of events cannot be discovered by observations of events *within* that chain. Biochemist Jacque Monod wrote, "Chance, blind but free, [lies] at the very root of the stupendous edifice of evolution," and that is true, but it tells us nothing. The flow of a gas through a pipe, looked at in detail, is also governed by chance — by innumerable random collisions between molecules — and yet the body of the gas flows on through the pipe, driven by the pressure of the piston. Just so, evolving populations "flow" in the direction of the pressures exerted by the environment, although, when looked at in detail, all within the population seems random variation. Random events provide the raw materials for evolution according to Darwin, but they do not rule the process. Evolution is governed by the forces of the physical environment. Whether these forces, in turn, are governed by a larger force — whether all we see around us is the unfolding of some larger plan — are questions science cannot answer.

Robert Jastrow
JUNE 4, 1984

1

Autobiography

[*In the last years of his life, Darwin wrote his autobiography,
more as a personal sketch for his descendants than as a public
document. As autobiographies go, his is as candid as most. It
illuminates not only his formative experiences, but also, in its
final section, gives his own assessment of his particular intellectual gifts. I find especially touching the expression of these latter views, where — in the idiom of Victorian virtue — he attributes his extraordinary success in science mainly to hard work
and persistence.*

*The apparently limited extent of Darwin's formal scientific
training deserves comment. Through the first part of the nineteenth century, the British universities in general and Oxford
and Cambridge in particular placed little value on the sciences.
There was no science curriculum as such, and mathematics and
classics provided the only avenues to a degree at Cambridge.
Even most who held professorships in the sciences regarded their
responsibilities lightly, seldom presenting more than a lecture or
two each year (and poorly attended ones at that). The seat of
scientific activity at Cambridge was, therefore, extracurricular,
often formed around local societies and, more informally, around
evening meetings of interested faculty and undergraduates. Thus
Darwin credits his association in this manner with the botanist, John Henslow, as the most singularly influential aspect of
his scientific education at Cambridge. By the standards of the
day, Darwin was probably as well prepared as any neophyte
scientist.*]

May 31st, 1876
Recollections of the Development of my mind and character

A GERMAN EDITOR having written to me to ask for an account of the development of my mind and character with some sketch of my autobiography, I have thought that the attempt would amuse me, and might possibly interest my children or their children. I know that it would have interested me greatly to have read even so short and dull a sketch of the mind of my grandfather written by himself, and what he thought and did and how he worked. I have attempted to write the following account of myself, as if I were a dead man in another world looking back at my own life. Nor have I found this difficult, for life is nearly over with me. I have taken no pains about my style of writing.

I was born at Shrewsbury on February 12th, 1809. I have heard my Father say that he believed that persons with powerful minds generally had memories extending far back to a very early period of life. This is not my case, for my earliest recollection goes back only to when I was a few months over four years old, when we went to near Abergele for sea-bathing, and I recollect some events and places there with some little distinctness.

My mother died in July 1817, when I was a little over eight years old, and it is odd that I can remember hardly anything about her except her death-bed, her black velvet gown, and her curiously constructed work-table. I believe that my forgetfulness is partly due to my sisters, owing to their great grief, never being able to speak about her or mention her name; and partly to her previous invalid state. In the spring of this same year I was sent to a day-school in Shrewsbury, where I staid a year. Before going to school I was educated by my sister Caroline, but I doubt whether this plan answered. I have been told that I was much slower in learning than my younger sister Catherine, and I believe that I was in many ways a naughty boy. Caroline was extemely kind, clever and zealous; but she was too zealous in trying to improve me; for I clearly remember after this long interval of years, saying to myself when about to enter a room where she was — "What will she blame me for now?" and I made myself dogged so as not to care what she might say.

By the time I went to this day-school my taste for natural history, and more expecially for collecting, was well developed. I tried to make out the names of plants, and collected all sorts of things, shells, seals, franks, coins, and minerals. The passion for collecting, which leads a man to be a systematic naturalist, a virtuoso or a miser, was very strong in me, and was clearly innate, as none of my sisters or brother ever had this taste.

Looking back as well as I can at my character during my school life, the only qualities which at this period promised well for the future, were, that I had strong and diversified tastes, much zeal for whatever interested me, and a keen pleasure in understanding any complex subject or thing. I was taught Euclid by a private tutor, and I distinctly remember the intense satisfaction which the clear geometrical proofs gave me. I remember with equal distinctness the delight which my uncle gave me (the father of Francis Galton) by explaining the principle of the vernier of a barometer. With respect to diversified tastes, independently of science, I was fond of reading various books, and I used to sit for hours reading the historical plays of Shakespeare, generally in an old window in the thick walls of the school. I read also other poetry, such as the recently published poems of Byron, Scott, and Thomson's *Seasons*. I mention this because later in life I wholly lost, to my great regret, all pleasure from poetry of any kind, including Shakespeare. In connection with pleasure from poetry I may add that in 1822 a vivid delight in scenery was first awakened in my mind, during a riding tour on the borders of Wales, and which has lasted longer than any other aesthetic pleasure.

⁂

With respect to science, I continued collecting minerals with much zeal, but quite unscientifically — all that I cared for was a new *named* mineral, and I hardly attempted to classify them. I must have observed insects with some little care, for when ten years old (1819) I went for three weeks to Plas Edwards on the sea-coast in Wales, I was very much interested and surprised at seeing a large black and scarlet Hemipterous insect, many moths (Zygæna) and a Cicindela, which are not found in Shropshire. I almost made up my mind to begin collecting all the insects which I could find dead, for on consulting my sister, I concluded that it was not right to kill insects for the sake of making a collection. From reading White's *Selborne* I took much pleasure in watching the habits of birds, and even made notes on the subject. In my simplicity I remember wondering why every gentleman did not become an ornithologist.

Towards the close of my school life, my brother worked hard at chemistry and made a fair laboratory with proper apparatus in the tool-house in the garden, and I was allowed to aid him as a servant in most of his experiments. He made all the gases and many compounds, and I read with care several books on chemistry, such as Henry and Parkes' *Chemical Catechism*. The subject interested me greatly, and we often used to go on working till rather late at night. This was the best part of my education at school, for it showed me practically the meaning of experimental science. The fact that we worked at chemistry somehow got known at school, and as it was an unprecedented fact, I was nicknamed "Gas." I was also once publicly rebuked by the head-master, Dr. Butler, for thus wasting my time over such useless subjects; and he called me very unjustly a "poco curante," and as I did not understand what he meant it seemed to me a fearful reproach.

As I was doing no good at school, my father wisely took me away at a rather earlier age than usual, and sent me (October 1825) to Edinburgh University with my brother, where I stayed for two years or sessions. My brother was completing his medical studies, though I do not believe he ever really intended to practise, and I was sent there to commence them. But soon after this period I became convinced from various small circumstances that my father would leave me property enough to subsist on with some comfort, though I never imagined that I should be so rich a man as I am; but my belief was sufficient to check any strenuous effort to learn medicine.

The instruction at Edinburgh was altogether by Lectures, and these were intolerably dull, with the exception of those on chemistry by Hope; but to my mind there are no advantages and many disadvantages in lectures compared with reading. Dr. Duncan's lectures on Materia Medica at 8 o'clock on a winter's morning are something fearful to remember. Dr. Munro made his lectures on human anatomy as dull as he was himself, and the subject disgusted me. It has proved one of the greatest evils in my life that I was not urged to practise dissection, for I should soon have got over my disgust; and the practice would have been invaluable for all my future work. This has been an irremediable evil, as well as my incapacity to draw. I also attended regularly the clinical wards in the Hospital. Some of the cases distressed me a good deal, and I still have vivid pictures before me of some of them; but I was not so foolish as to allow this to lessen my attendance. I cannot understand why this part of my medical course did not interest me in a greater degree; for during the summer before coming to Edinburgh I began attending some of the poor people, chiefly children and women in Shrewsbury: I wrote down as full an account as I could of the cases with all the symptoms, and read them aloud to my father, who suggested further enquiries, and advised me what medicines to give, which I made up myself. At one time I had at least a dozen patients, and I felt a keen interest in the work. My father, who was by far the best judge of character whom I ever knew, declared that I should make a successful physician, — meaning by this, one who got many patients. He maintained that the chief element of success was exciting confidence; but what he saw in me which convinced him that I should create confidence I know not. I also attended on two occasions the operating theatre in the hospital at Edinburgh, and saw two very bad operations, one on a child, but I rushed away before they were completed. Nor did I ever attend again, for hardly any inducement would have been strong enough to make me do so; this being long before the blessed days of chloroform. The two cases fairly haunted me for many a long year.

My Brother staid only one year at the University, so that during the second year I was left to my own resources; and this was an advantage, for I became well acquainted with several young men fond of natural science. One of these was Ainsworth, who afterwards published his travels in Assyria: he was a Wernerian geologist and knew a little about many subjects, but was superficial and very glib with his tongue. Dr. Coldstream was a very different young man, prim, formal, highly religious and most kind-hearted: he afterwards published some good zoological articles. A third young man was Hardie, who would I think have made a good botanist, but died early in India. Lastly, Dr. Grant, my senior by several years, but how I became acquainted with him I cannot remember: he published some first-rate zoological papers, but after coming to London as Professor in University College, he did nothing more in science — a fact which has always been inexplicable to me. I knew him well; he was dry and formal in manner, but with much enthusiasm beneath this outer crust. He one day, when we were walking together, burst forth in high admiration of Lamarck and his views on evolution. I listened in silent astonishment, and as far as I can judge, without any effect on my mind. I had previously read the *Zoönomia* of my grandfather, in which similar views are maintained, but without producing any effect on me. Nevertheless it is probable that the hearing rather early in life such views maintained and praised may have favoured my upholding them under a different form in my *Origin of Species*. At this time I admired greatly the *Zoönomia*; but on reading it a second time after an interval of ten or fifteen years, I was much disappointed, the proportion of speculation being so large to the facts given.

Drs. Grant and Coldstream attended much to marine Zoology, and I often accompanied the former to collect animals in the tidal pools, which I dissected as well as I could. I also became friends with some of the Newhaven fishermen, and sometimes accompanied them when they trawled for oysters, and thus got many specimens. But from not having had any regular practice in dissection, and from possessing only a wretched microscope my attempts were very poor. Nevertheless I made one interesting little discovery, and read about the beginning of the year 1826, a short paper on the subject before the Plinian Socy. This was that the so-called ova of Flustra had the power of independent movement by means of cilia, and were in fact larvae. In another short paper I showed that little globular bodies which had been supposed to be the young state of *Fucus loreus* were the egg-cases of the worm-like *Pontobdella muricata*.

Mr. Leonard Horner also took me once to a meeting of the Royal Society of Edinburgh, where I saw Sir Walter Scott in the chair as President, and he apologised to the meeting as not feeling fitted for such a position. I looked at him and at the whole scene with some awe and reverence; and I think it was owing to this visit during my youth and to my having attended the Royal Medical Society, that I felt the honour of being elected a few years ago an honorary member of both these Societies, more than any other similar honour. If I had been told at that time that I should one day have been thus honoured, I declare that I should have thought it as ridiculous and improbable as if I had been told that I should be elected King of England.

Cambridge, 1828–1831

AFTER HAVING spent two sessions in Edinburgh, my father perceived or he heard from my sisters, that I did not like the thought of being a physician, so he proposed that I should become a clergyman. He was very properly vehement against my turning an idle sporting man, which then seemed my probable destination. I asked for some time to consider, as from what little I had heard and thought on the subject I had scruples about declaring my belief in all the dogmas of the Church of England; though otherwise I liked the thought of being a country clergyman. Accordingly I read with care *Pearson on the Creed* and a few other books on divinity; and as I did not then in the least doubt the strict and literal truth of every word in the Bible, I soon persuaded myself that our Creed must be fully accepted. It never struck me how illogical it was to say that I believed in what I could not understand and what is in fact unintelligible. I might have said with entire truth that I had no wish to dispute any dogma; but I never was such a fool as to feel and say 'credo quia incredibile.'

Considering how fiercely I have been attacked by the orthodox it seems ludicrous that I once intended to be a clergyman. Nor was this intention and my father's wish ever formally given up, but died a natural death when on leaving Cambridge I joined the *Beagle* as Naturalist. If the phrenologists are to be trusted, I was well fitted in one respect to be a clergyman. A few years ago the Secretaries of a German psychological society asked me earnestly by letter for a photograph of myself; and some time afterwards I received the proceedings of one of the meetings in which it seemed that the shape of my head had been the subject of a public discussion, and one of the speakers declared that I had the bump of Reverence developed enough for ten Priests.

As it was decided that I should be a clergyman, it was necessary that I should go to one of the English universities and take a degree; but as I had never opened a classical book since leaving school, I found to my dismay that in the two intervening years I had actually forgotten, incredible as it may appear, almost everything which I had learnt even to some few of the Greek letters. I did not therefore proceed to Cambridge at the usual time in October, but worked with a private tutor in Shrewsbury and went to Cambridge after the Christmas vacation, early in 1828. I soon recovered my school standard of knowledge, and could translate easy Greek books, such as Homer and the Greek Testament with moderate facility.

During the three years which I spent at Cambridge my time was wasted, as far as the academical studies were concerned, as completely as at Edinburgh and at school. I attempted mathematics, and even went during the summer of 1828 with a private tutor (a very dull man) to Barmouth, but I got on very slowly. The work was repugnant to me, chiefly from my not being able to see any meaning in the early steps in algebra. This impatience was very foolish, and in after years I have deeply regretted that I did not proceed far enough at least to understand something of the great leading principles of mathematics; for men thus endowed seem to have an extra sense. But I do not believe that I should ever have succeeded beyond a very low grade. With respect to Classics I did nothing except attend a few compulsory college lectures, and the attendance was almost nominal. In my second year I had to work for a month or two to pass the Little Go, which I did easily. Again in my last year I worked with some earnestness for my final degree of B.A., and brushed up my Classics together with a little Algebra and Euclid, which latter gave me much pleasure, as it did whilst at school. In order to pass the B.A. examination, it was, also, necessary to get up Paley's *Evidences of Christianity*, and his *Moral Philosophy*. This was done in a thorough manner, and I am convinced that I could have written out the whole of the *Evidences* with perfect correctness, but not of course in the clear language of Paley. The logic of this book and as I may add of his *Natural Theology* gave me as much delight as did Euclid. The careful study of these works, without attempting to learn any part by rote, was the only part of the Academical Course which, as I then felt and as I still believe, was of the least use to me in the education of my mind. I did not at that time trouble myself about Paley's premises; and taking these on trust I was charmed and convinced by the long line of argumentation. By answering well the examination questions in Paley, by doing Euclid well, and by not failing miserably in Classics, I gained a good place among the δc πολλοί, or crowd of men who do not go in for honours. Oddly enough I cannot remember how high I stood, and my memory fluctuates between the fifth, tenth, or twelfth name on the list.

Although as we shall presently see there were some redeeming features in my life at Cambridge, my time was sadly wasted there and worse than wasted. From my passion for shooting and for hunting and, when this failed, for riding across country I got into a sporting set, including some dissipated lowminded young men. We used often to dine together in the evening, though these dinners often included men of a higher stamp, and we sometimes drank too much, with jolly singing and playing at cards afterwards. I know that I ought to feel ashamed of days and evenings thus spent, but as some of my friends were very pleasant and we were all in the highest spirits, I cannot help looking back to these times with much pleasure.

But no pursuit at Cambridge was followed with nearly so much eagerness or gave me so much pleasure as collecting beetles. It was the mere passion for collecting, for I did not dissect them and rarely compared their external characters with published descriptions, but got them named anyhow. I will give a proof of my zeal: one day, on tearing off some old bark, I saw two rare beetles and seized one in each hand; then I saw a third and new kind, which I could not bear to lose, so that I popped the one which I held in my right hand into my mouth. Alas it ejected some intensely acrid fluid, which burnt my tongue so that I was forced to spit the beetle out, which was lost, as well as the third one.

I have not as yet mentioned a circumstance which influenced my whole career more than any other. This was my friendship with Prof. Henslow. Before coming up to Cambridge, I had heard of him from my brother as a man who knew every branch of science, and I was accordingly prepared to reverence him. He kept open house once every week, where all undergraduates and several older members of the University, who were attached to science, used to meet in the evening. I soon got, through Fox, an invitation, and went there regularly. Before long I became well acquainted with Henslow, and during the latter half of my time at Cambridge took long walks with him on most days; so that I was called by some of the dons "the man who walks with Henslow"; and in the evening I was very often asked to join his family dinner. His knowledge was great in botany, entomology, chemistry, mineralogy, and geology. His strongest taste was to draw conclusions from long-continued minute observations. His judgment was excellent, and his whole mind well-balanced; but I do not suppose that anyone would say that he possessed much original genius.

As I had at first come up to Cambridge at Christmas, I was forced to keep two terms after passing my final examination, at the commencement of 1831; and Henslow then persuaded me to begin the study of geology. Therefore on my return to Shropshire I examined sections and coloured a map of parts round Shrewsbury. Professor Sedgwick intended to visit N. Wales in the beginning of August to pursue his famous geological investigation amongst the older rocks, and Henslow asked him to allow me to accompany him. Accordingly he came and slept at my Father's house.

A short conversation with him during this evening produced a strong impression on my mind. Whilst examining an old gravel-pit near Shrewsbury a labourer told me that he had found in it a large worn tropical Volute shell, such as may be seen on the chimney-pieces of cottages; and as he would not sell the shell I was convinced that he had really found it in the pit. I told Sedgwick of the fact, and he at once said (no doubt truly) that it must have been thrown away by someone into the pit; but then added, if really embedded there it would be the greatest misfortune to geology, as it would overthrow all that we know about the superficial deposits of the midland counties. These gravel-beds belonged in fact to the glacial period, and in after years I found in them broken arctic shells. But I was then utterly astonished at Sedgwick not being delighted at so wonderful a fact as a tropical shell being found near the surface in the middle of England. Nothing before had ever made me thoroughly realise, though I had read various scientific books, that science consists in grouping facts so that general laws or conclusions may be drawn from them.

Next morning we started for Llangollen, Conway, Bangor, and Capel Curig. This tour was of decided use in teaching me a little how to make out the geology of a country. Sedgwick often sent me on a line parallel to his, telling me to bring back specimens of the rocks and to mark the stratification on a map. I have little doubt that he did this for my good, as I was too ignorant to have aided him. On this tour I had a striking instance how easy it is to overlook phenomena, however conspicuous, before they have been observed by anyone. We spent many hours in Cwm Idwal, examining all the rocks with extreme care, as Sedgwick was anxious to find fossils in them; but neither of us saw a trace of the wonderful glacial phenomena all around us; we did not notice the plainly scored rocks, the perched boulders, the lateral and terminal moraines. Yet these phenomena are so conspicuous that, as I declared in a paper published many years afterwards in the *Philosophical Magazine*, a house burnt down by fire did not tell its story more plainly than did this valley. If it had still been filled by a glacier, the phenomena would have been less distinct than they now are.

At Capel Curig I left Sedgwick and went in a straight line by compass and map across the mountains to Barmouth, never following any track unless it coincided with my course. I thus came on some strange wild places and enjoyed much this manner of travelling. I visited Barmouth to see some Cambridge friends who were reading there, and thence returned to Shrewsbury and to Maer for shooting; for at that time I should have thought myself mad to give up the first days of partridge-shooting for geology or any other science.

[Maer was home to Darwin's "Uncle Jos," Josiah Wedgewood, the son of the founder of the celebrated pottery.]

Voyage of the 'Beagle': from Dec. 27, 1831 to Oct. 2, 1836

ON RETURNING home from my short geological tour in N. Wales, I found a letter from Henslow, informing me that Captain Fitz-Roy was willing to give up part of his own cabin to any young man who would volunteer to go with him without pay as naturalist to the Voyage of the *Beagle*. I have given as I believe in my M.S. Journal an account of all the circumstances which then occurred; I will here only say that I was instantly eager to accept the offer, but my father strongly objected, adding the words fortunate for me, — "If you can find any man of common sense, who advises you to go, I will give my consent." So I wrote that evening and refused the offer. On the next morning I went to Maer to be ready for September 1st, and whilst out shooting, my uncle sent for me, offering to drive me over to Shrewsbury and talk with my father. As my uncle thought it would be wise in me to accept the offer, and as my father always maintained that he was one of the most sensible men in the world, he at once consented in the kindest manner. I had been rather extravagant at Cambridge and to console my father said, "that I should be deuced clever to spend more than my allowance whilst on board the *Beagle*"; but he answered with a smile, "But they all tell me you are very clever."

Next day I started for Cambridge to see Henslow, and thence to London to see Fitz-Roy, and all was soon arranged. Afterwards on becoming very intimate with Fitz-Roy, I heard that I had run a very narrow risk of being rejected, on account of the shape of my nose! He was an ardent disciple of Lavater, and was convinced that he could judge a man's character by the outline of his features; and he doubted whether anyone with my nose could possess sufficient energy and determination for the voyage. But I think he was afterwards well-satisfied that my nose had spoken falsely.

Fitz-Roy's character was a singular one, with many very noble features: he was devoted to his duty, generous to a fault, bold, determined, indomitably energetic, and an ardent friend to all under his sway. He would undertake any sort of trouble to assist those whom he thought deserved assistance. He was a handsome man, strikingly like a gentleman, with highly courteous manners, which resembled those of his maternal uncle, the famous Lord Castlereagh, as I was told by the Minister at Rio.

Fitz-Roy's temper was a most unfortunate one. This was shown not only by passion but by fits of long-continued moroseness against those who had offended him. His temper was usually worst in the early morning, and with his eagle eye he could generally detect something amiss about the ship, and was then unsparing in his blame. The junior officers when they relieved each other in the forenoon used to ask "whether much hot coffee had been served out this morning, — " which meant how was the Captain's temper? He was also somewhat suspicious and occasionally in very low spirits, on one occasion bordering on insanity. He seemed to me often to fail in sound judgment or common sense. He was extremely kind to me, but was a man very difficult to live with on the intimate terms which necessarily followed from our messing by ourselves in the same cabin. We had several quarrels; for when out of temper he was utterly unreasonable. For instance, early in the voyage at Bahia in Brazil he defended and praised slavery, which I abominated, and told me that he had just visited a great slave-owner, who had called up many of his slaves and asked them whether they were happy, and whether they wished to be free, and all answered "No." I then asked him, perhaps with a sneer, whether he thought that the answers of slaves in the presence of their master was worth anything. This made him excessively angry, and he said that as I doubted his word, we could not live any longer together. I thought that I should have been compelled to leave the ship; but as soon as the news spread, which it did quickly, as the captain sent for the first lieutenant to assuage his anger by abusing me, I was deeply gratified by receiving an invitation from all the gun-room officers to mess with them. But after a few hours Fitz-Roy showed his usual magnanimity by sending an officer to me with an apology and a request that I would continue to live with him. I remember another instance of his candour. At Plymouth before we sailed, he was extremely angry with a dealer in crockery who refused to exchange some article purchased in his shop: the Captain asked the man the price of a very expensive set of china and said "I should have purchased this if you had not been so disobliging." As I knew that the cabin was amply stocked with crockery, I doubted whether he had any such intention; and I must have shown my doubts in my face, for I said not a word. After leaving the shop he looked at me, saying You do not believe what I have said, and I was forced to own that it was so. He was silent for a few minutes and then said You are right, and I acted wrongly in my anger at the blackguard.

At Conception in Chile, poor Fitz-Roy was sadly overworked and in very low spirits; he complained bitterly to me that he must give a great party to all the inhabitants of the place. I remonstrated and said that I could see no such necessity on his part under the circumstances. He then burst out into a fury, declaring that I was the sort of man who would receive any favours and make no return. I got up and left the cabin without saying a word, and returned to Conception where I was then lodging. After a few days I came back to the ship and was received by the Captain as cordially as ever, for the storm had by that time quite blown over. The first Lieutenant, however, said to me: "Confound you, philosopher, I wish you would not quarrel with the skipper; the day you left the ship I was dead-tired (the ship was refitting) and he kept me walking the deck till midnight abusing you all the time." The difficulty of living on good terms with a Captain of a Man-of-War is much increased by its being almost mutinous to answer him as one would answer anyone else; and by the awe in which he is held — or was held in my time, by all on board. I remember hearing a curious instance of this in the case of the purser of the *Adventure*, — the ship which sailed with the *Beagle* during the first voyage. The Purser was in a store in Rio de Janeiro, purchasing rum for the ship's company, and a little gentleman in plain clothes walked in. The Purser said to him, "Now Sir, be so kind as to taste this rum, and give me your opinion of it." The gentleman did as he was asked, and soon left the store. The store-keeper then asked the Purser whether he knew that he had been speaking to the Captain of a Line of Battleships which had just come into the harbour. The poor Purser was struck dumb with horror; he let the glass of spirit drop from his hand onto the floor, and immediately went on board, and no persuasion, as an officer on the *Adventure* assured me, could make him go on shore again for fear of meeting the Captain after his dreadful act of familiarity.

I saw Fitz-Roy only occasionally after our return home, for I was always afraid of unintentionally offending him, and did so once, almost beyond mutual reconciliation. He was afterwards very indignant with me for having published so unorthodox a book (for he became very religious) as the *Origin of Species*. Towards the close of his life he was as I fear, much impoverished, and this was largely due to his generosity. Anyhow after his death a subscription was raised to pay his debts. His end was a melancholy one, namely suicide, exactly like that of his uncle Ld. Castlereagh, whom he resembled closely in manner and appearance.

His character was in several respects one of the most noble which I have ever known, though tarnished by grave blemishes.

The voyage of the *Beagle* has been by far the most important event in my life and has determined my whole career; yet it depended on so small a circumstance as my uncle offering to drive me 30 miles to Shrewsbury, which few uncles would have done, and on such a trifle as the shape of my nose. I have always felt that I owe to the voyage the first real training or education of my mind. I was led to attend closely to several branches of natural history, and thus my powers of observation were improved, though they were already fairly developed.

The investigation of the geology of all the places visited was far more important, as reasoning here comes into play. On first examining a new district nothing can appear more hopeless than the chaos of rocks; but by recording the stratification and nature of the rocks and fossils at many points, always reasoning and predicting what will be found elsewhere, light soon begins to dawn on the district, and the structure of the whole becomes more or less intelligible. I had brought with me the first volume of Lyell's *Principles of Geology*, which I studied attentively; and this book was of the highest service to me in many ways. The very first place which I examined, namely St. Jago in the Cape Verde islands, showed me clearly the wonderful superiority of Lyell's manner of treating geology, compared with that of any other author, whose works I had with me or ever afterwards read.

I need not here refer to the events of the voyage — where we went and what we did — as I have given a sufficiently full account in my published Journal. The glories of the vegetation of the Tropics rise before my mind at the present time more vividly than anything else. Though the sense of sublimity, which the great deserts of Patagonia and the forest-clad mountains of

Tierra del Fuego excited in me, has left an indelible impression on my mind. The sight of a naked savage in his native land is an event which can never be forgotten. Many of my excursions on horseback through wild countries, or in the boats, some of which lasted several weeks, were deeply interesting; their discomfort and some degree of danger were at that time hardly a drawback and none at all afterwards. I also reflect with high satisfaction on some of my scientific work, such as solving the problem of coral-islands, and making out the geological structure of certain islands, for instance, St. Helena. Nor must I pass over the discovery of the singular relations of the animals and plants inhabiting the several islands of the Galapagos archipelago, and of all of them to the inhabitants of South America.

As far as I can judge of myself I worked to the utmost during the voyage from the mere pleasure of investigation, and from my strong desire to add a few facts to the great mass of facts in natural science. But I was also ambitious to take a fair place among scientific men, — whether more ambitious or less so than most of my fellow-workers I can form no opinion.

Towards the close of our voyage I received a letter whilst at Ascension, in which my sisters told me that Sedgwick had called on my father and said that I should take a place among the leading scientific men. I could not at the time understand how he could have learnt anything of my proceedings, but I heard (I believe afterwards) that Henslow had read some of the letters which I wrote to him before the Philosophical Soc. of Cambrige and had printed them for private distribution. My collection of fossil bones, which had been sent to Henslow, also excited considerable attention amongst palæontologists. After reading this letter I clambered over the mountains of Ascension with a bounding step and made the volcanic rocks resound under my geological hammer! All this shows how ambitious I was; but I think that I can say with truth that in after years, though I cared in the highest degree for the approbation of such men as Lyell and Hooker, who were my friends, I did not care much about the general public. I do not mean to say that a favourable review or a large sale of my books did not please me greatly; but the pleasure was a fleeting one, and I am sure that I have never turned one inch out of my course to gain fame.

[Darwin returned to England in October 1836. After a short time in Cambridge he settled in London, where he remained for over five years. There he plunged into his career as a geologist, becoming a protégé of Sir Charles Lyell. Although his early influence in geology, under Robert Jameson at Edinburgh and Adam Sedgwick at Cambridge, had been more in the tradition of catastrophism (by which previous geological processes are held to consist primarily in episodic flooding and volcanism of no longer experienced intensity), his observations in South America persuaded him of the strength of Lyell's uniformitarian views (according to which the geological record is an accumulation of ordinary and observable natural processes acting over a very long time). His most important contribution to geology, The Structure and Distribution of Coral Reefs (1842), was an ingenious demonstration of Lyellian thinking: coral reefs result from gradual subsidence of midoceanic volcanic cones upon which attached coral become slowly deposited in a subsurface zone; the present varieties in form of such reefs represent different stages in a continuous process of formation. Darwin's success as a geologist assured him already a "fair place among scientific men." Serving as secretary of The Geological Society, he was well incorporated into the British scientific community even before he expanded his career as a naturalist.

In 1839 Darwin married Emma Wedgwood, the daughter of Uncle Josiah. He was fortunate to have received settlements and *bequests from his father, Robert, a successful physician; these, along with Emma's dowry, provided income ample to pursue his profession without financial distraction. Only in matters of health was Darwin's future clouded. In South America he seems to have contracted Chagas' disease, a parasitic infection related to African sleeping sickness, and its symptoms began to recur shortly after his return. For the balance of his life he was plagued with intermittent and often protracted sieges of debilitating illness, requiring carefully measured apportionment of his physical energies to complete his work.* Equally distressing, his children also revealed signs of infirmity; only seven of ten survived to adulthood, and several who did were chronically ill.*

In 1842, perhaps to conserve his resources by escaping the social demands of London, Darwin moved his growing family to Down, in the Kentish countryside. In a large and comfortable house he maintained there a relatively secluded and always methodical existence for most of the forty years following.]

I have now mentioned all the books which I have published, and these have been the milestones in my life, so that little remains to be said. I am not conscious of any change in my mind during the last thirty years, excepting in one point presently to be mentioned; nor indeed could any change have been expected unless one of general deterioration. But my father lived to his eighty-third year with his mind as lively as ever it was, and all his faculties undimmed; and I hope that I may die before my mind fails to a sensible extent. I think that I have become a little more skilful in guessing right explanations and in devising experimental tests; but this may probably be the result of mere practice, and of a larger store of knowledge. I have as much difficulty as ever in expressing myself clearly and con-

---

* Chagas' disease may not be entirely to blame for Darwin's invalidism. A variety of medical diagnoses, including psychomatic disorders, have been considered by Colp.[1]

cisely; and this difficulty has caused me a very great loss of time; but it has had the compensating advantage of forcing me to think long and intently about every sentence, and thus I have been often led to see errors in reasoning and in my own observations or those of others.

There seems to be a sort of fatality in my mind leading me to put at first my statement and proposition in a wrong or awkward form. Formerly I used to think about my sentences before writing them down; but for several years I have found that it saves time to scribble in a vile hand whole pages as quickly as I possibly can, contracting half the words; and then correct deliberately. Sentences thus scribbled down are often better ones than I could have written deliberately.

My books have sold largely in England, have been translated into many languages, and passed through several editions in foreign countries. I have heard it said that the success of a work abroad is the best test of its enduring value. I doubt whether this is at all trustworthy; but judged by this standard my name ought to last for a few years. Therefore it may be worth while for me to try to analyse the mental qualities and the conditions on which my success has depended; though I am aware that no man can do this correctly.

I have no great quickness of apprehension or wit which is so remarkable in some clever men, for instance Huxley. I am therefore a poor critic: a paper or book, when first read, generally excites my admiration, and it is only after considerable reflection that I perceive the weak points. My power to follow a long and purely abstract train of thought is very limited; I should, moreover, never have succeeded with metaphysics or mathematics. My memory is extensive, yet hazy: it suffices to make me cautious by vaguely telling me that I have observed or read something opposed to the conclusion which I am drawing, or on the other hand in favour of it; and after a time

I can generally recollect where to search for my authority. So poor in one sense is my memory, that I have never been able to remember for more than a few days a single date or a line of poetry.

Some of my critics have said, "Oh, he is a good observer, but has no power of reasoning." I do not think that this can be true, for the *Origin of Species* is one long argument from the beginning to the end, and it has convinced not a few able men. No one could have written it without having some power of reasoning. I have a fair share of invention and of common sense or judgment, such as every fairly successful lawyer or doctor must have, but not I believe, in any higher degree.

On the favourable side of the balance, I think that I am superior to the common run of men in noticing things which easily escape attention, and in observing them carefully. My industry has been nearly as great as it could have been in the observation and collection of facts. What is far more important, my love of natural science has been steady and ardent. This pure love has, however, been much aided by the ambition to be esteemed by my fellow naturalists. From my early youth I have had the strongest desire to understand or explain whatever I observed, — that is, to group all facts under some general laws. These causes combined have given me the patience to reflect or ponder for any number of years over any unexplained problem. As far as I can judge, I am not apt to follow blindly the lead of other men. I have steadily endeavoured to keep my mind free, so as to give up any hypothesis, however much beloved (and I cannot resist forming one on every subject), as soon as facts are shown to be opposed to it. Indeed I have had no choice but to act in this manner, for with the exception of the Coral Reefs, I cannot remember a single first-formed hypothesis which had not after a time to be given up or greatly modified. This has naturally led me to distrust greatly deductive reasoning in the mixed sciences. On the other hand, I am not very sceptical, — a frame of mind which

I believe to be injurious to the progress of science; a good deal of scepticism in a scientific man is advisable to avoid much loss of time; for I have met with not a few men, who I feel sure have often thus been deterred from experiment or observations, which would have proved directly or indirectly serviceable.

In illustration, I will give the oddest case which I have known. A gentleman (who, as I afterwards heard, was a good local botanist) wrote to me from the Eastern counties that the seeds or beans of the common field-bean had this year everywhere grown on the wrong side of the pod. I wrote back, asking for further information, as I did not understand what was meant; but I did not receive any answer for a long time. I then saw in two newspapers, one published in Kent and the other in Yorkshire, paragraphs stating that it was a most remarkable fact that "the beans this year had all grown on the wrong side." So I thought that there must be some foundation for so general a statement. Accordingly, I went to my gardener, an old Kentish man, and asked him whether he had heard anything about it; and he answered, "Oh, no, Sir, it must be a mistake, for the beans grow on the wrong side only on Leap-year, and this is not Leap-year." I then asked him how they grew on common years and how on leap-years, but soon found out that he knew absolutely nothing of how they grew at any time; but he stuck to his belief.

After a time I heard from my first informant, who, with many apologies, said that he should not have written to me had he not heard the statement from several intelligent farmers; but that he had since spoken again to every one of them, and not one knew in the least what he had himself meant. So that here a belief — if indeed a statement with no definite idea attached to it can be called a belief — had spread over almost the whole of England without any vestige of evidence.

My habits are methodical, and this has been of not a little use for my particular line of work. Lastly, I have had ample leisure from not having to earn my own bread. Even ill-health, though it has annihilated several years of my life, has saved me from the distractions of society and amusement.

Therefore, my success as a man of science, whatever this may have amounted to, has been determined, as far as I can judge, by complex and diversified mental qualities and conditions. Of these the most important have been — the love of science — unbounded patience in long reflecting over any subject — industry in observing and collecting facts — and a fair share of invention as well as of common-sense. With such moderate abilities as I possess, it is truly surprising that thus I should have influenced to a considerable extent the beliefs of scientific men on some important points.

August 3rd 1876

2

The Voyage of the Beagle

[In 1835 the Beagle arrived in the Galapagos Islands, on the homeward leg of its 1832–1836 voyage. It remained there for five weeks, only three of which Darwin spent ashore. Notwithstanding its brevity, however, almost certainly there was no experience more pivotal to his nascent ideas on evolution, for he witnessed what he came later to regard as nature's laboratory. I shall not elaborate here the evolutionary significance of his observations, saving this for Chapter XII of the Origin. Suffice it to note that it is extremely probable that he did not awake to the fact — let alone the means — of evolution until his return to England, and when he did he dropped hardly a hint in the Voyage of the Beagle, compiled later from his travel diaries.[1] Yet this chapter contains virtually all the evidence wanted to adduce evolution; thus I leave open to the reader the opportunity for its independent rediscovery.]

CHAPTER XVII

Galapagos Archipelago

September 15th. — This archipelago consists of ten principal islands, of which five exceed the others in size. They are situated under the Equator, and between five and six hundred miles westward of the coast of America. They are all formed of volcanic rocks; a few fragments of granite curiously glazed and altered by the heat can hardly be considered as an exception. Some of the craters, surmounting the larger islands, are of immense size, and they rise to a height of between three and four thousand feet. Their flanks are studded by innumerable smaller orifices. I scarcely hesitate to affirm, that there must be in the whole archipelago at least two thousand craters. These consist either of lava and scoriæ, or of finely-stratified, sandstone-like tuff. Most of the latter are beautifully symmetrical; they owe their origin to eruptions of volcanic mud without any lava: it is a remarkable circumstance that every one of the twenty-

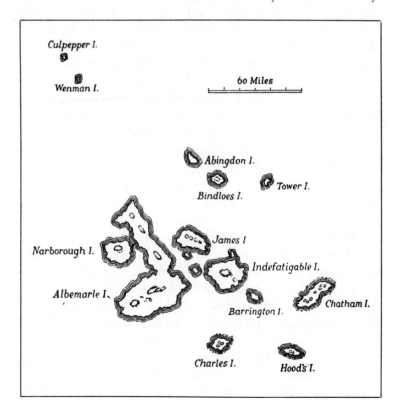

Culpepper I.

Wenman I.

60 Miles

Abingdon I.

Bindloes I.

Tower I.

Narborough I.

James I

Indefatigable I.

Albemarle I.

Barrington I.

Chatham I.

Charles I.

Hood's I.

eight tuff-craters which were examined had their southern sides either much lower than the other sides, or quite broken down and removed. As all these craters have apparently been formed when standing in the sea, and as the waves from the trade-wind and the swell from the open Pacific here unite their forces on the southern coasts of all the islands, this singular uniformity in the broken state of the craters, composed of the soft and yielding tuff, is easily explained.

Considering that these islands are placed directly under the equator, the climate is far from being excessively hot; this seems chiefly caused by the singularly low temperature of the surrounding water, brought here by the great southern Polar current. Excepting during one short season, very little rain falls, and even then it is irregular; but the clouds generally hang low. Hence, whilst the lower parts of the island are very sterile, the upper parts, at a height of a thousand feet and upwards, possess a damp climate and a tolerably luxuriant vegetation. This is especially the case on the windward sides of the islands, which first receive and condense the moisture from the atmosphere.

In the morning (17th) we landed on Chatham Island, which, like the others, rises with a tame and rounded outline, broken here and there by scattered hillocks, the remains of former craters. Nothing could be less inviting than the first appearance. A broken field of black basaltic lava, thrown into the most rugged waves, and crossed by great fissures, is every where covered by stunted, sunburnt brushwood, which shows little signs of life. The dry and parched surface, being heated by the noonday sun, gave to the air a close and sultry feeling, like that from a stove: we fancied even that the bushes smelt unpleasantly. Although I diligently tried to collect as many plants as possible, I succeeded in getting very few; and such wretched-looking little weeds would have better become an arctic than an equatorial Flora. The brushwood appears, from a short distance, as leafless as our trees during winter; and it was some time before I discovered that not only almost every plant was now in full leaf, but that the greater number were in flower. The commonest bush is one of the Euphorbiaceæ: an acacia and a great odd-looking cactus are the only trees which afford any shade. After the season of heavy rains, the islands are said to appear for a short time partially green. The volcanic island of Fernando Noronha, placed in many respects under nearly similar conditions, is the only other country where I have seen a vegetation at all like this of the Galapagos islands.

23rd. — The *Beagle* proceeded to Charles Island. This archipelago has long been frequented, first by the Bucaniers, and latterly by whalers, but it is only within the last six years, that a small colony has been established here. The inhabitants are between two and three hundred in number: they are nearly all people of colour, who have been banished for political crimes from the Republic of the Equator, of which Quito is the capital. The settlement is placed about four and a half miles inland, and at a height probably of a thousand feet. In the first part of the road we passed through leafless thickets, as in Chatham Island. Higher up, the woods gradually became greener: and as soon as we crossed the ridge of the island, we were cooled by a fine southerly breeze, and our sight refreshed by a green and thriving vegetation. In this upper region coarse grasses and ferns abound; but there are no tree-ferns: I saw nowhere any member of the Palm family, which is the more singular, as 360 miles northward, Cocos Island takes its name from the number of cocoa-nuts. The houses are irregularly scattered over a flat space of ground, which is cultivated with sweet potatoes and bananas. It will not easily be imagined how pleasant the sight of black mud was to us, after having been so long accustomed to the parched soil of Peru and northern Chile. The inhabitants, although complaining of poverty, obtain, without much trouble, the means of subsistence. In the woods there are many wild pigs and goats; but the staple article of animal food is supplied by the tortoises. Their numbers have of course been greatly reduced in this island, but the people yet count on two days' hunting giving them food for the rest of the week. It is said that formerly single vessels have taken away as many as seven hundred, and that the ship's company of a frigate some years since brought down in one day two hundred tortoises to the beach.

September 29th. — We doubled the south-west extremity of Albemarle Island, and the next day were nearly becalmed between it and Narborough Island. Both are covered with immense deluges of black naked lava, which have flowed either over the rims of the great caldrons, like pitch over the rim of a pot in which it has been boiled, or have burst forth from smaller orifices on the flanks; in their descent they have spread over miles of the sea-coast. On both of these islands, eruptions are known to have taken place; and in Albemarle, we saw a small jet of smoke curling from the summit of one of the great craters. In the evening we anchored in Bank's Cove, in Albemarle Island. The next morning I went out walking. To the south of the broken tuff-crater, in which the *Beagle* was anchored, there was another beautifully symmetrical one of an elliptic form; its longer axis was a little less than a mile, and its depth about 500 feet. At its bottom there was a shallow lake, in the middle of which a tiny crater formed an islet. The day was overpoweringly hot, and the lake looked clear and blue: I hurried down the cindery slope, and choked with dust eagerly tasted the water — but, to my sorrow, I found it salt as brine.

The rocks on the coast abounded with great black lizards, between three and four feet long; and on the hills, an ugly yellowish-brown species was equally common. We saw many of this latter kind, some clumsily running out of our way, and others shuffling into their burrows. I shall presently describe in more detail the habits of both these reptiles. The whole of this northern part of Albemarle Island is miserably sterile.

October 8th. — We arrived at James Island; this island, as well as Charles Island, were long since thus named after our kings of the Stuart line. Mr. Bynoe, myself, and our servants were left here for a week, with provisions and a tent, whilst the *Beagle* went for water. We found here a party of Spaniards, who had been sent from Charles Island to dry fish, and to salt tortoise-meat. About six miles inland, and at the height of nearly 2000 feet, a hovel had been built in which two men lived, who were employed in catching tortoises, whilst the others were fishing on the coast. I paid this party two visits, and slept there one night. As in the other islands, the lower region was covered by nearly leafless bushes, but the trees were here of a larger growth than elsewhere, several being two feet and some even two feet nine inches in diameter. The upper region being kept damp by the clouds, supports a green and flourishing vegetation. So damp was the ground, that there were large beds of a coarse cyperus, in which great numbers of a very small water-rail lived and bred. While staying in this upper region, we lived entirely upon tortoise-meat: the breastplate roasted (as the Gauchos do *carne con cuero*), with the flesh on it, is very good; and the young tortoises make excellent soup; but otherwise the meat to my taste is indifferent.

One day we accompanied a party of the Spaniards in their whale-boat to a salina, or lake from which salt is procured. After landing, we had a very rough walk over a rugged field of recent lava, which has almost surrounded a tuff-crater, at the bottom of which the salt-lake lies. The water is only three or four inches deep, and rests on a layer of beautifully crystallized, white salt. The lake is quite circular, and is fringed with a border of bright green succulent plants; the almost precipitous walls of the crater are clothed with wood, so that the scene was altogether both picturesque and curious. A few years since, the sailors belonging to a sealing-vessel murdered their captain in this quiet spot; and we saw his skull lying among the bushes.

The natural history of these islands is eminently curious, and well deserves attention. Most of the organic productions are aboriginal creations, found nowhere else; there is even a difference between the inhabitants of the different islands; yet all show a marked relationship with those of America, though separated from that continent by an open space of ocean, between 500 and 600 miles in width. The archipelago is a little world within itself, or rather a satellite attached to America, whence it has derived a few stray colonists, and has received the general character of its indigenous productions. Considering the small size of these islands, we feel the more astonished at the number of their aboriginal beings, and at their confined range. Seeing every height crowned with its crater, and the boundaries of most of the lava-streams still distinct, we are led to believe that within a period, geologically recent, the unbroken ocean was here spread out. Hence, both in space and time, we seem to be brought somewhat near to that great fact — that mystery of mysteries — the first appearance of new beings on this earth.

Of terrestrial mammals, there is only one which must be considered as indigenous, namely, a mouse (Mus Galapagoensis), and this is confined, as far as I could ascertain, to Chatham Island, the most easterly island of the group. It belongs, as I am informed by Mr. Waterhouse, to a division of the family of mice characteristic of America. At James Island, there is a rat sufficiently distinct from the common kind to have been named and described by Mr. Waterhouse; but as it belongs to the old-world division of the family, and as this island has been frequented by ships for the last hundred and fifty years, I can hardly doubt that this rat is merely a variety, produced by the new and peculiar climate, food, and soil, to which it has been subjected. Although no one has a right to speculate without distinct facts, yet even with respect to the Chatham Island mouse, it should be borne in mind, that it may possibly be an American species imported here; for I have seen, in a most unfrequented part of the Pampas, a native mouse living in the roof of a newly-built hovel, and therefore its transportation in a vessel is not improbable: analogous facts have been observed by Dr. Richardson in North America.

Of land-birds I obtained twenty-six kinds; all peculiar to the group and found nowhere else, with the exception of one lark-like finch from North America (Dolichonyx oryzivorus), which ranges on that continent as far north as 54°, and generally frequents marshes. The other twenty-five birds consist, firstly, of a hawk, curiously intermediate in structure between a Buzzard and the American group of carrion-feeding Polybori; and with these latter birds it agrees most closely in every habit and even tone of voice. Secondly, there are two owls, representing the short-eared and white barn-owls of Europe. Thirdly, a wren, three tyrant fly-catchers (two of them species of Pyrocephalus, one or both of which would be ranked by some ornithologists as only varieties), and a dove — all analogous to, but distinct from, American species. Fourthly, a swallow, which though differing from the Progne purpurea of both Americas, only in being rather duller coloured, smaller, and slenderer, is considered by Mr. Gould as specifically distinct. Fifthly, there are three species of mocking-thrush — a form highly characteristic of America. The remaining land-birds form a most singular group of finches, related to each other in the structure of their beaks, short tails, form of body, and plumage: there are thirteen species, which Mr. Gould has divided into four sub-groups. All these species are peculiar to this archipelago; and so is the whole group, with the exception of one species of the sub-group Cactornis, lately brought from Bow Island, in the Low Archipelago. Of Cactornis, the two species may be often seen climbing about the flowers of the great cactus-trees; but all the other species of this group of finches, mingled together in flocks, feed on the dry and sterile ground of the lower districts. The males of all, or certainly of the greater number, are jet black; and the females (with perhaps one or two exceptions) are brown. The most curious fact is the perfect gradation in the size of the beaks in the different species of Geospiza, from one as large as that of a hawfinch to that of a chaffinch, and (if Mr. Gould is right in including his sub-group, Certhidea, in the main group), even to that of a warbler. The largest beak in the genus Geospiza is shown in Fig. 1, and the smallest in Fig. 3; but instead of their being only one intermediate species, with a beak of the size shown in Fig. 2, there are no less than six species with insensibly graduated beaks. The beak of the sub-group Certhidea is shown in Fig. 4. The beak of Cactornis is somewhat like that of a starling; and that of the fourth sub-group, Camarhynchus, is slightly parrot-shaped. Seeing this gradation and diversity of structure in one small, intimately related group of birds, one might really fancy that from an original paucity of birds in this archipelago, one species had been taken and modified for dif-

ferent ends. In a like manner it might be fancied that a bird originally a buzzard had been induced here to undertake the office of the carrion-feeding Polybori of the American continent.

Of waders and water-birds I was able to get only eleven kinds, and of these only three (including a rail confined to the damp summits of the islands) are new species. Considering the wandering habits of the gulls, I was surprised to find that the species inhabiting these islands is peculiar, but allied to one from the southern parts of South America. The far greater peculiarity of the land-birds, namely, twenty-five out of twenty-six being new species or at least new races, compared with the waders and web-footed birds, is in accordance with the greater range which these latter orders have in all parts of the world. We shall hereafter see this law of aquatic forms, whether marine or fresh-water, being less peculiar at any given point of the earth's surface than the terrestrial forms of the same classes, strikingly illustrated in the shells, and in a lesser degree in the insects of this archipelago.

We will now turn to the order of reptiles, which gives the most striking character to the zoology of these islands. The species are not numerous, but the numbers of individuals of each species are extraordinarily great. There is one small lizard belonging to a South American genus, and two species (and probably more) of the Amblyrhynchus — a genus confined to the Galapagos Islands. There is one snake which is numerous; it is identical, as I am informed by M. Bibron, with the Psammophis Temminckii from Chile. Of sea-turtle I believe there is more than one species; and of tortoises there are, as we shall presently show, two or three species or races. Of toads and frogs there are none: I was surprised at this, considering how well suited for them the temperate and damp upper woods appeared to be. It recalled to my mind the remark made by Bory St.

Vincent, namely, that none of this family are found on any of the volcanic islands in the great oceans. As far as I can ascertain from various works, this seems to hold good throughout the Pacific, and even in the large islands of the Sandwich archipelago. . . . The absence of the frog family in the oceanic islands is the more remarkable, when contrasted with the case of lizards, which swarm on most of the smallest islands. May this difference not be caused by the greater facility with which the eggs of lizards, protected by calcareous shells, might be transported through salt-water, than could the slimy spawn of frogs?

I will first describe the habits of the tortoise (Testudo nigra, formerly called Indica), which has been so frequently alluded to. These animals are found, I believe, on all the islands of the Archipelago; certainly on the greater number. They frequent in preference the high damp parts, but they likewise live in the lower and arid districts. I have already shown, from the numbers which have been caught in a single day, how very numerous they must be. Some grow to an immense size: Mr. Lawson, an Englishman, and vice-governor of the colony, told us that he had seen several so large, that it required six or eight men to lift them from the ground; and that some had afforded as much as two hundred pounds of meat. The old males are the largest, the females rarely growing to so great a size: the male can readily be distinguished from the female by the greater length of its tail. The tortoises which live on those islands where there is no water, or in the lower and arid parts of the others, feed chiefly on the succulent cactus. Those which frequent the higher and damp regions eat the leaves of various trees, a kind of berry (called guayavita) which is acid and austere, and likewise a pale green filamentous lichen (Usnera plicata), that hangs in tresses from the boughs of the trees.

The tortoise is very fond of water, drinking large quantities, and wallowing in the mud. The larger islands alone possess springs, and these are always situated towards the central parts, and at a considerable height. The tortoises, therefore, which frequent the lower districts, when thirsty, are obliged to travel from a long distance. Hence broad and well-beaten paths branch off in every direction from the wells down to the sea-coast; and the Spaniards by following them up, first discovered the watering-places. When I landed at Chatham Island, I could not imagine what animal travelled so methodically along well-chosen tracks. Near the springs it was a curious spectacle to behold many of these huge creatures, one set eagerly travelling onwards with outstretched necks, and another set returning, after having drunk their fill. When the tortoise arrives at the spring, quite regardless of any spectator, he buries his head in the water above his eyes, and greedily swallows great mouthfuls, at the rate of about ten in a minute. The inhabitants say each animal stays three or four days in the neighbourhood of the water, and then returns to the lower country; but they differed respecting the frequency of these visits. The animal probably regulates them according to the nature of the food on which it has lived. It is, however, certain, that tortoises can subsist even on those islands, where there is no other water than what falls during a few rainy days in the year.

The tortoises, when purposely moving towards any point, travel by night and day, and arrive at their journey's end much sooner than would be expected. The inhabitants, from observing marked individuals, consider that they travel a distance of about eight miles in two or three days. One large tortoise, which I watched, walked at the rate of sixty yards in ten minutes, that is 360 yards in the hour, or four miles a day, — allowing a little time for it to eat on the road. During the breeding season, when the male and female are together, the male utters a hoarse roar or bellowing, which, it is said, can be heard at the distance of more than a hundred yards. The female never uses her voice, and the male only at these times; so that when the people hear this noise, they know that the two are together. They were at this time (October) laying their eggs. The female, where the soil is sandy, deposits them together, and covers them up with sand; but where the ground is rocky she drops them indiscriminately in any hole: Mr. Bynoe found seven placed in a fissure. The egg is white and spherical; one which I measured was seven inches and three-eighths in circumference, and therefore larger than a hen's egg. The young tortoises, as soon as they are hatched, fall a prey in great numbers to the carrion-feeding buzzard. The old ones seem generally to die from accidents, as from falling down precipices: at least, several of the inhabitants told me, that they had never found one dead without some evident cause.

There can be little doubt that this tortoise is an aboriginal inhabitant of the Galapagos; for it is found on all, or nearly all, the islands, even on some of the smaller ones where there is no water; had it been an imported species, this would hardly have been the case in a group which has been so little frequented. Moreover, the old Bucaniers found this tortoise in greater numbers even than at present: Wood and Rogers also, in 1708, say that it is the opinion of the Spaniards, that it is found nowhere else in this quarter of the world. It is now widely distributed; but it may be questioned whether it is in any other place an aboriginal. The bones of a tortoise at Mauritius, associated with those of the extinct Dodo, have generally been considered as belonging to this tortoise: if this had been so, undoubtedly it must have been there indigenous; but M. Bibron informs me that he believes that it was distinct, as the species now living there certainly is.

The Amblyrhynchus, a remarkable genus of lizards, is confined to this archipelago: there are two species, resembling each other in general form, one being terrestrial and the other aquatic. This latter species (A. cristatus) was first characterized by Mr.

Bell, who well foresaw, from its short, broad head, and strong claws of equal length, that its habits of life would turn out very peculiar, and different from those of its nearest ally, the Iguana. It is extremely common on all the islands throughout the group, and lives exclusively on the rocky sea-beaches, being never found, at least I never saw one, even ten yards in-shore. It is a hideous-looking creature, of a dirty black colour, stupid, and sluggish in its movements. The usual length of a full-grown one is about a yard, but there are some even four feet long; a large one weighed twenty pounds: on the island of Albemarle they seem to grow to a greater size than elsewhere. Their tails are flattened sideways, and all four feet partially webbed. They are occasionally seen some hundred yards from the shore, swimming about; and Captain Collnett, in his Voyage, says, "They go to sea in herds a-fishing, and sun themselves on the rocks; and may be called alligators in miniature." It must not, however, be supposed that they live on fish. When in the water this lizard swims with perfect ease and quickness, by a serpentine movement of its body and flattened tail — the legs being motionless and closely collapsed on its sides. A seaman on board sank one, with a heavy weight attached to it, thinking thus to kill it directly; but when, an hour afterwards, he drew up the line, it was quite active. Their limbs and strong claws are admirably adapted for crawling over the rugged and fissured masses of lava, which everywhere form the coast. In such situations, a group of six or seven of these hideous reptiles may oftentimes be seen on the black rocks, a few feet above the surf, basking in the sun with outstretched legs.

I opened the stomachs of several, and found them largely distended with minced sea-weed (Ulvæ), which grows in thin foliaceous expansions of a bright green or a dull red colour. I do not recollect having observed this sea-weed in any quantity on the tidal rocks; and I have reason to believe it grows at the bottom of the sea, at some little distance from the coast. If such be the case, the object of these animals occasionally going out to sea is explained. The stomach contained nothing but the sea-weed. Mr. Bynoe, however, found a piece of a crab in one; but this might have got in accidentally, in the same manner as I have seen a caterpillar, in the midst of some lichen, in the paunch of a tortoise. The intestines were large, as in other herbivorous animals. The nature of this lizard's food, as well as the structure of its tail and feet, and the fact of its having been seen voluntarily swimming out at sea, absolutely prove its aquatic habits; yet there is in this respect one strange anomaly, namely, that when frightened it will not enter the water. Hence it is easy to drive these lizards down to any little point overhanging the sea, where they will sooner allow a person to catch hold of their tails than jump into the water. They do not seem to have any notion of biting; but when much frightened they squirt a drop of fluid from each nostril. I threw one several times as far as I could, into a deep pool left by the retiring tide; but it invariably returned in a direct line to the spot where I stood. It swam near the bottom, with a very graceful and rapid movement, and occasionally aided itself over the uneven ground with its feet. As soon as it arrived near the edge, but still being under water, it tried to conceal itself in the tufts of sea-weed, or it entered some crevice. As soon as it thought the danger was past, it crawled out on the dry rocks, and shuffled away as quickly as it could. I several times caught this same lizard, by driving it down to a point, and though possessed of such perfect powers of diving and swimming, nothing would induce it to enter the water; and as often as I threw it in, it returned in the manner above described. Perhaps this singular piece of apparent stupidity may be accounted for by the circumstance, that this reptile has no enemy whatever on shore, whereas at sea it must often fall a prey to the numerous sharks. Hence, probably, urged by a fixed and hereditary instinct that the shore is its place of safety, whatever the emergency may be, it there takes refuge.

We will now turn to the terrestrial species (A. Demarlii), with a round tail, and toes without webs. This lizard, instead of being found like the other on all the islands, is confined to the central part of the archipelago, namely to Albemarle, James, Barrington, and Indefatigable islands. To the southward, in Charles, Hood, and Chatham islands, and to the northward, in Towers, Bindloes, and Abingdon, I neither saw nor heard of any. It would appear as if it had been created in the centre of the archipelago, and thence had been dispersed only to a certain distance. Some of these lizards inhabit the high and damp parts of the islands, but they are much more numerous in the lower and sterile districts near the coast. I cannot give a more forcible proof of their numbers, than by stating that when we were left at James Island, we could not for some time find a spot free from their burrows on which to pitch our single tent. Like their brothers the sea-kind, they are ugly animals, of a yellowish orange beneath, and of a brownish red colour above: from their low facial angle they have a singularly stupid appearance. They are, perhaps, of a rather less size than the marine species; but several of them weighed between ten and fifteen pounds. In their movements, they are lazy and half torpid. When not frightened, they slowly crawl along with their tails and bellies dragging on the ground. They often stop, and doze for a minute or two, with closed eyes and hind legs spread out on the parched soil.

They inhabit burrows, which they sometimes make between fragments of lava, but more generally on level patches of the soft sandstone-like tuff. The holes do not appear to be very deep, and they enter the ground at a small angle; so that when walking over these lizard-warrens, the soil is constantly giving way, much to the annoyance of the tired walker. This animal, when making its burrow, works alternately the opposite sides of its body. One front leg for a short time scratches up the soil, and throws it towards the hind foot, which is well placed so as to heave it beyond the mouth of the hole. That side of the body being tired, the other takes up the task, and so on alternately. I watched one for a long time, till half its body was buried; I then walked up and pulled it by the tail; at this it was greatly astonished, and soon shuffled up to see what was the matter; and then stared me in the face, as much as to say, "What made you pull my tail?"

I opened the stomachs of several, and found them full of vegetable fibres and leaves of different trees, especially of an acacia. In the upper region they live chiefly on the acid and astringent berries of the guayavita, under which trees I have seen these lizards and huge tortoises feeding together. To obtain the acacia-leaves they crawl up the low stunted trees; and it is not uncommon to see a pair quietly browsing, whilst seated on a branch several feet above the ground. These lizards, when cooked, yield a white meat, which is liked by those whose stomachs soar above all prejudices. Humboldt has remarked that in intertropical South America, all lizards which inhabit dry regions are esteemed delicacies for the table. The inhabitants state that those which inhabit the upper damp parts drink water, but that the others do not, like the tortoises, travel up for it from the lower sterile country. At the time of our visit, the females had within their bodies numerous large, elongated eggs, which they lay in their burrows: the inhabitants seek them for food.

These two species of Amblyrhynchus agree, as I have already stated, in their general structure, and in many of their habits. Neither have that rapid movement, so characteristic of the genera Lacerta and Iguana. They are both herbivorous, although the kind of vegetation on which they feed is so very different. Mr. Bell has given the name to the genus from the shortness of the snout; indeed, the form of the mouth may almost be compared to that of the tortoise: one is led to suppose that this is an adaptation to their herbivorous appetites. It is very interesting thus to find a well-characterized genus, having its marine and terrestrial species, belonging to so confined a portion of the world. The aquatic species is by far the most remarkable, because it is the only existing lizard which lives on marine vegetable productions. As I at first observed, these islands are not so remarkable for the number of the species of reptiles, as for that of the individuals; when we remember the well-beaten paths made by the thousands of huge tortoises — the many turtles — the great warrens of the terrestrial Amblyrhynchus — and the groups of the marine species basking on the coast-rocks of every island — we must admit that there is no other quarter of the world where this Order replaces the herbivorous mammalia in so extraordinary a manner. The geologist on hearing this will probably refer back in his mind to the Secondary epochs, when lizards, some herbivorous, some carnivorous, and of dimensions comparable only with our existing whales, swarmed on the land and in the sea. It is, therefore, worthy of his observation, that this archipelago, instead of possessing a humid climate and rank vegetation, cannot be considered otherwise than extremely arid, and, for an equatorial region, remarkably temperate.

The botany of this group is fully as interesting as the zoology, Dr. J. Hooker will soon publish in the "Linnean Transactions" a full account of the Flora, and I am much indebted to him for the following details. Of flowering plants there are, as far as at present is known, 185 species, and 40 cryptogamic species, making together 225; of this number I was fortunate enough to bring home 193. Of the flowering plants, 100 are new species, and are probably confined to this archipelago. Dr. Hooker conceives that, of the plants not so confined, at least 10 species found near the cultivated ground at Charles Island, have been imported. It is, I think, surprising that more American species have not been introduced naturally, considering that the distance is only between 500 and 600 miles from the continent; and that (according to Collnett, p. 58) drift-wood, bamboos, canes, and the nuts of a palm, are often washed on the south-eastern shores. The proportion of 100 flowering plants out of 185 (or 175 excluding the imported weeds) being new, is sufficient, I conceive, to make the Galapagos Archipelago a distinct botanical province; but this Flora is not nearly so peculiar as that of St. Helena, nor, as I am informed by Dr. Hooker, of Juan Fernandez. The peculiarity of the Galapageian Flora is best shown in certain families; — thus there are 21 species of Compositæ, of which 20 are peculiar to this archipelago; these belong to twelve genera, and of these genera no less than ten are confined to the archipelago! Dr. Hooker informs me that the Flora has an undoubted Western American character; nor can he detect in it any affinity with that of the Pacific. If, therefore, we except the eighteen marine, the one fresh-water, and one land-shell, which have apparently come here as colonists from the central islands of the Pacific, and likewise the one distinct Pacific species of the Galapageian group of finches, we see that this archipelago, though standing in the Pacific Ocean, is zoologically part of America.

If this character were owing merely to immigrants from America, there would be little remarkable in it; but we see that a vast majority of all the land animals, and that more than half of the flowering plants, are aboriginal productions. It was most striking to be surrounded by new birds, new reptiles, new shells, new insects, new plants, and yet by innumerable trifling details of structure, and even by the tones of voice and plumage of the birds, to have the temperate plains of Patagonia, or the hot dry deserts of Northern Chile, vividly brought before my eyes. Why, on these small points of land, which within a late geological period must have been covered by the ocean, which are formed of basaltic lava, and therefore differ in geological character from the American continent, and which are placed under a peculiar climate, — why were their aboriginal inhabitants, associated, I may add, in different proportions both in kind and number from those on the continent, and therefore acting on each other in a different manner — why were they created on American types of organization? It is probable that the islands of the Cape de Verd group resemble, in all their physical conditions, far more closely the Galapagos Islands than these latter physically resemble the coast of America; yet the aboriginal inhabitants of the two groups are totally unlike; those of the Cape de Verd Islands bearing the impress of Africa, as the inhabitants of the Galapagos Archipelago are stamped with that of America.

I have not as yet noticed by far the most remarkable feature in the natural history of this archipelago; it is, that the different islands to a considerable extent are inhabited by a different set of beings. My attention was first called to this fact by the Vice-Governor, Mr. Lawson, declaring that the tortoises differed from the different islands, and that he could with certainty tell from which island any one was brought. I did not for some time pay sufficient attention to this statement, and I had already partially mingled together the collections from two of the islands. I never dreamed that islands, about fifty or sixty miles apart, and most of them in sight of each other, formed of precisely the same rocks, placed under a quite similar climate, rising to a nearly equal height, would have been differently tenanted; but we shall soon see that this is the case. It is the fate of most voyagers, no sooner to discover what is most interesting in any locality, than they are hurried from it; but I ought, perhaps, to be thankful that I obtained sufficient material to establish this most remarkable fact in the distribution of organic beings.

The inhabitants, as I have said, state that they can distinguish the tortoises from the different islands; and that they differ not only in size, but in other characters. Captain Porter has described those from Charles and from the nearest island to it, namely, Hood Island, as having their shells in front thick and turned up like a Spanish saddle, whilst the tortoises from James Island are rounder, blacker, and have a better taste when cooked. M. Bibron, moreover, informs me that he has seen what he considers two distinct species of tortoise from the Galapagos, but he does not know from which islands. The specimens that I brought from three islands were young ones; and probably owing to this cause, neither Mr. Gray nor myself could find in them any specific differences. I have remarked that the marine Amblyrhynchus was larger at Albemarle Island than elsewhere; and M. Bibron informs me that he has seen two distinct aquatic species of this genus; so that the different islands probably have their representative species or races of the Amblyrhynchus, as well as of the tortoise. My attention was first thoroughly aroused, by comparing together the numerous specimens, shot by myself and several other parties on board, of the mocking-thrushes, when, to my astonishment, I discovered that all those from Charles Island belonged to one species (Mimus trifasciatus); all from Albemarle Island to M. parvulus; and all from James and Chatham Islands (between which two other islands are situated, as connecting links) belonged to M. melanotis. These two latter species are closely allied, and would by some ornithologists be considered as only well-marked races or varieties; but the Mimus trifasciatus is very distinct. Unfortunately most of the specimens of the finch tribe were mingled together; but I have strong reasons to suspect that some of the species of the sub-group Geospiza are confined to separate islands. If the different islands have their representatives of Geospiza, it may help to explain the singularly large number of the species of this sub-group in this one small archipelago, and as a probable consequence of their numbers, the perfectly graduated series in the size of their beaks. Two species of the sub-group Cactornis, and two of Camarhynchus, were procured in the archipelago; and of the numerous specimens of these two sub-groups shot by four collectors at James Island, all were found to belong to one species of each; whereas the numerous specimens shot either on Chatham or Charles Island (for the two sets were mingled together) all belonged to the two other species: hence we may feel almost sure that these islands possess· their representative species of these two sub-groups. In land-shells this law of distribution does not appear to hold good. In my very small collection of insects, Mr. Waterhouse remarks, that of those which were ticketed with their locality, not one was common to any two of the islands.

If we now turn to the Flora, we shall find the aboriginal plants of the different islands wonderfully different. I give all the following results on the high authority of my friend Dr. J. Hooker. I may premise that I indiscriminately collected everything in flower on the different islands, and fortunately kept my collections separate. Too much confidence, however, must not be placed in the proportional results, as the small collections brought home by some other naturalists, though in some respects confirming the results, plainly show that much remains to be done in the botany of this group: the Leguminosæ, moreover, have as yet been only approximately worked out: —

Name of Island.	Total No. of Species.	No. of Species found in other parts of the world.	No. of Species confined to the Galapagos Archipelago.	No. confined to the one Island.	No. of Species confined to the Galapagos Archipelago, but found on more than the one Island.
James Island	71	33	38	30	8
Albemarle Island	46	18	26	22	4
Chatham Island	32	16	16	12	4
Charles Island	68	39 (or 29, if the probably imported plants be subtracted)	29	21	8

Hence we have the truly wonderful fact, that in James Island, of the thirty-eight Galapageian plants, or those found in no other part of the world, thirty are exclusively confined to this one island; and in Albemarle Island, of the twenty-six aboriginal Galapageian plants, twenty-two are confined to this one island, that is, only four are at present known to grow in the other islands of the archipelago; and so on, as shown in the above table, with the plants from Chatham and Charles Islands.

The distribution of the tenants of this archipelago would not be nearly so wonderful, if, for instance, one island had a mocking-thrush, and a second island some other quite distinct genus; — if one island had its genus of lizard, and a second island another distinct genus, or none whatever; — or if the different islands were inhabited, not by representative species of the same genera of plants, but by totally different genera, as does to a certain extent hold good; for, to give one instance, a large berry-bearing tree at James Island has no representative species in Charles Island. But it is the circumstance, that several of the islands possess their own species of the tortoise, mocking-thrush, finches, and numerous plants, these species having the same general habits, occupying analogous situations, and obviously filling the same place in the natural economy of this archipelago, that strikes me with wonder. It may be suspected that some of these representative species, at least in the case of the tortoise and of some of the birds, may hereafter prove to be only well-marked races; but this would be of equally great interest to the philosophical naturalist. I have said that most of the islands are in sight of each other: I may specify that Charles Island is fifty miles from the nearest part of Chatham Island, and thirty-three miles from the nearest part of Albemarle Island. Chatham Island is sixty miles from the nearest part of James Island, but there are two intermediate islands between them which were not visited by me. James Island is only ten miles from the nearest part of Albemarle Island, but the two points where the collections were made are thirty-two miles apart. I must repeat, that neither the nature of the soil, nor height of the land, nor the climate, nor the general character of the associated beings, and therefore their action one on another, can differ much in the different islands. If there be any sensible difference in their climates, it must be between the windward group (namely Charles and Chatham Islands), and that to leeward; but there seems to be no corresponding difference in the productions of these two halves of the archipelago.

The only light which I can throw on this remarkable difference in the inhabitants of the different islands, is, that very strong currents of the sea running in a westerly and W.N.W. direction must separate, as far as transportal by the sea is concerned, the southern islands from the northern ones; and between these northern islands a strong N.W. current was observed, which must effectually separate James and Albemarle Islands. As the archipelago is free to a most remarkable degree from gales of wind, neither the birds, insects, nor lighter seeds would be blown from island to island. And lastly, the profound depth of the ocean between the islands, and their apparently recent (in a geological sense) volcanic origin, render it highly unlikely that they were ever united; and this, probably, is a far more important consideration than any other, with respect to the geographical distribution of their inhabitants. Reviewing the facts here given, one is astonished at the amount of creative force, if such an expression may be used, displayed on these small, barren, and rocky islands; and still more so, at its diverse yet analogous action on points so near each other. I have said that the Galapagos Archipelago might be called a satellite attached to America, but it should rather be called a group of satellites, physically similar, organically distinct, yet intimately related to each other, and all related in a marked, though much lesser degree, to the great American continent.

3

The Lengthy Delay

It is one of the many ironies of Darwin's career that what we esteem as the centerpiece of his voluminous scholarship — *On the Origin of Species* — he wrote in thirteen months, planning it only as an introduction to some grander, future exposition. Instead, it was this hurriedly prepared "abstract," as he called it, that became over the course of its six editions the durable war-horse from which he was to parry his critics and to advance his views for so long as he campaigned. We shall begin this introduction to the *Origin* with something of the circumstances leading up to its publication, for the urgency of its preparation hardly belies its slow incubation over the twenty years preceding. We know that in 1837, a year after his return to England, Darwin began the first of his "transmutation notebooks" in which he started to compile evidence and musings on the fact and means of evolution. By 1842 his efforts had progressed sufficiently to permit the writing of a short preliminary draft, supplanted by a more substantial version of 230 pages in 1844.[1] The latter essay in particular was so close in spirit and form to the *Origin* of 1859 that Darwin's almost exclusive devotion during the long interim to the tasks of authoring first a volume on the geology of South America and then a monumental four-volume series on barnacles may appear somewhat curious. So the writing of the *Origin* poses at the outset the question of the paradoxical coupling of the delay and haste that attended its creation.

Let us first take up the matter of delay. Beyond regretting in retrospect the eight years consumed by barnacles, Darwin offers very little help on this point:

The Cirripedes form a highly varying and difficult group of species to class; and my work was of considerable use to me, when
I had to discuss in the *Origin of Species* the principle of a natural classification. Nevertheless, I doubt whether the work was
worth the consumption of so much time.[2]

But comparing the *Origin* to the essay of 1844, it is plain that
the *Origin* received only the smallest benefit of this side-trip
into descriptive zoology. We need a better explanation for this
lengthy intermission.

As we might have imagined, the forces that stayed Darwin's
hand are to be found in the climate of scientific opinion that
then prevailed. There had been published prior to the *Origin*
a scattered rash of claims for organic evolution, and these
without exception had been derisively greeted by the Victorian
scientific community in England. Indeed, Darwin's grandfather, Erasmus Darwin, had before the turn of that century
published such a work. Better remembered by our own century is the evolutionary theory of Lamarck, who posited a continual and recurrent progression of organisms up an ascending
chain-of-being; ascent in this case owed to inner needs arising
in response to environmental flux, leading in turn to inherited
change. But the disparagement with which British science held
up these (and various other such) accounts was little in comparison to the scorn it reserved for *Vestiges of the Natural History of Creation*, a treatise published anonymously in 1844.[3]
Its author, a respected Scottish publisher named Robert
Chambers (whose identity was revealed only much later), postulated the operation of grandiose cosmological principles, not
the least being a succession of species progressing from "lower"
to "higher" in accord with divinely ordained natural principles
tending toward perfection. Among organisms, man's was said
to be the highest estate, but opportunities for further advancement were held out. This hardly original promise undoubtedly
contributed to the book's immense popular success, its enthu-

siasts numbering many prominent figures from outside scientific circles.

If by now you wonder why Darwin (along with Alfred Russel Wallace, whom we shall soon meet) should be granted priority for discovery of organic evolution, then a moment's digression is wanted. All antecedent mechanisms for evolutionary change hinged in one fashion or another upon some metaphysical property — vitalist strivings, impulses toward perfection, and so forth — as to a considerable extent they were also teleological, invoking some transcendental tendency to arrive at a preestablished goal. Darwin's entire account, by contrast, was materialist, founded exclusively upon empirically grounded phenomena. Strictly, then, his was the first scientific explanation for evolution.

No doubt the attacks on *Vestiges* by Darwin's colleagues were inspired as much by its favorable public reception as by the bad science that certainly it purveyed, but a deeper vein of disapproval stemmed from its having threatened the tenuous accord between science and religion. Early in that century Paley's *Natural Theology* had provided the proper interpretive scheme for the natural sciences through the argument for divine creation from design:

> There cannot be a design without a designer; contrivance, without a contriver; order without choice; arrangement, without anything capable of arranging; subserviency and relation to a purpose, without that which could intend a purpose; means suitable to an end, and executing their office in accomplishing that end, without the end ever having been contemplated, or the means accommodated to it. Arrangement, disposition of parts, subserviency of means to an end, relation of instruments to a use, imply the presence of intelligence and mind.[4]

If this has a familiar ring, it is to Paley that we owe the analogy of the watchmaker and the watch, by which — unmindful of an illogical confounding of origin, function, and purpose —

persistent adherents of special creation still invite us to adduce the origins of the exquisite complexity of a perfectly designed eye. During the 1830s the argument from design had been strongly buttressed by the Bridgewater Treatises, a series of works explicitly commissioned to reveal the plan of God in nature, whose contributors included those most prominent in British science. While this perspective emphasized the order of the organic world, it was determinedly averse to a naturalistic explanation for changes in this order — most especially the progression of species. Chambers's donning of Paley's mantle, arguing in Vestiges that the cosmological laws he identified were statements of God's design, could not have been more impertinent.

Why the thesis of organic evolution was unwelcomed had less to do with its contradicting the tenets of revealed Christianity than with a set of broader and more diffuse concerns about the stable maintenance of society. To cite a single illustration, in his greatly influential Principles of Geology Sir Charles Lyell had already insisted that the formation of the earth's geological record was accomplished by ordinary climatological phenomena acting over vast periods, a view squarely in conflict with the biblicism of diluvian interpretation (holding all geological phenomena to be products of catastrophic events, such as the Flood). Yet Lyell's so-called uniformitarian explanations, if not universally upheld, were quite readily accommodated within the scope of scientific orthodoxy. And significantly, Lyell, who recognized in the fossils both the extinction and appearance of species, had brought his argument for gradual and continuous change of the earth's record to the threshold of organic progression but no further. The scientific community's distaste for evolution, in fact, stemmed as much as anything from a desire to keep fast the bonds of a moral and social order believed to mirror the natural order flowing from Providence. To repudiate divine creation of the organic world, and so to decouple man from God, was assuredly to invite civilization's demise. Species mutability simply had been that line that science had agreed not to cross. This will become plain enough when we examine the initial reactions to the *Origins*.

It ought to be fairly evident now why Darwin, with his nakedly materialistic theory of species origins, had been so long reluctant to press his case publicly. But why, when he finally chose otherwise, was he galvanized so suddenly into action? On the vigorous encouragement of the eminent Lyell, by now his confidant, Darwin began in 1856 to write up his theory once again. But two years later, only halfway completed, with a work already grown three or four times longer than the *Origin* was to be, he received a manuscript from Alfred Wallace, a naturalist working in Malaya, with a solicitation of his opinion. The manuscript was, of course, a close approximation to Darwin's own theory. After settling the matter of priority (Darwin yielded to Wallace, Wallace to Darwin, both agreed to become cofounders), a joint memoir was prepared and communicated to the Linnean Society by Lyell and Joseph Hooker, a distinguished botanist and close friend of Darwin. Their papers "excited very little attention," recalled Darwin in his *Autobiography*, "and the only published notice of them which I can remember was by Professor Haughton of Dublin, whose verdict was that all that was new in them was false, and what was true was old." [5] And so, to establish a prompt public claim on his theory, that next year he hastily abstracted the *Origin* from the longer and never-to-be-finished manuscript.

Before we come finally to the *Origin* itself, it is interesting to consider why a work of such colossal stature, however dutifully its reading is begun, is so rarely finished. In this, perhaps, it is akin to *Das Kapital*. Certainly the nub of its argument — that evolution is accomplished by natural selection operating on a source of inherited variation — is so readily transparent that everyone familiar with its rudiments can appreciate Huxley's oft-quoted "how extremely stupid not to have thought of that." And while the Victorian ornateness of its prose might contribute to its formidable first impression, I rather suspect that David Hull comes closer to the mark in noting that "the modern reader frequently grows impatient with Darwin's method in *Origin* of piling example upon example," going on to remark that this remains yet the format in evolutionary theory.[6] Without addressing here the methodological grounding of Darwin's theory (the interested reader will benefit from Hull's examination), it must be observed that in the *Origin* Darwin devised an entirely novel way of conducting scientific inquiry. Recognizing correctly that the acceptance of his theory perforce depended upon the number of observations harmoniously encompassed by it, he drew upon an extraordinary store of information to construct, as he put it, "one long argument from the beginning to the end."[7] His efforts ought not be displeasing.

4

On the Origin of Species

ON THE ORIGIN OF SPECIES

Introduction.

WHEN ON BOARD H.M.S. *Beagle*, as naturalist, I was much struck with certain facts in the distribution of the inhabitants of South America, and in the geological relations of the present to the past inhabitants of that continent. These facts seemed to me to throw some light on the origin of species — that mystery of mysteries, as it has been called by one of our greatest philosophers. On my return home, it occurred to me, in 1837, that something might perhaps be made out on this question by patiently accumulating and reflecting on all sorts of facts which could possibly have any bearing on it. After five years' work I allowed myself to speculate on the subject, and drew up some short notes; these I enlarged in 1844 into a sketch of the conclusions, which then seemed to me probable: from that period to the present day I have steadily pursued the same object. I hope that I may be excused for entering on these personal details, as I give them to show that I have not been hasty in coming to a decision.

My work is now nearly finished; but as it will take me two or three more years to complete it, and as my health is far from strong, I have been urged to publish this Abstract. I have more especially been induced to do this, as Mr. Wallace, who is now studying the natural history of the Malay archipelago, has arrived at almost exactly the same general conclusions that I have on the origin of species. Last year he sent to me a memoir on this subject, with a request that I would forward it to Sir Charles Lyell, who sent it to the Linnean Society, and it is published in the third volume of the Journal of that Society. Sir C. Lyell and Dr. Hooker, who both knew of my work — the latter having read my sketch of 1844 — honoured me by thinking it advisable to publish, with Mr. Wallace's excellent memoir, some brief extracts from my manuscripts.

This Abstract, which I now publish, must necessarily be imperfect. I cannot here give references and authorities for my several statements; and I must trust to the reader reposing some confidence in my accuracy. No doubt errors will have crept in, though I hope I have always been cautious in trusting to good authorities alone. I can here give only the general conclusions at which I have arrived, with a few facts in illustration, but which, I hope, in most cases will suffice. No one can feel more sensible than I do of the necessity of hereafter publishing in detail all the facts, with references, on which my conclusions have been grounded; and I hope in a future work to do this. For I am well aware that scarcely a single point is discussed in this volume on which facts cannot be adduced, often apparently leading to conclusions directly opposite to those at which I have arrived. A fair result can be obtained only by fully stating and balancing the facts and arguments on both sides of each question; and this cannot possibly be here done.

No one ought to feel surprise at much remaining as yet unexplained in regard to the origin of species and varieties, if he makes due allowance for our profound ignorance in regard to the mutual relations of all the beings which live around us. Who can explain why one species ranges widely and is very numerous, and why another allied species has a narrow range and is rare? Yet these relations are of the highest importance, for they determine the present welfare, and, as I believe, the future success and modification of every inhabitant of this world. Still less do we know of the mutual relations of the innumerable inhabitants of the world during the many past geological epochs in its history. Although much remains obscure, and will long remain obscure, I can entertain no doubt, after the most deliberate study and dispassionate judgment of which I am capable, that the view which most naturalists entertain, and which I formerly entertained — namely, that each species has been independently created — is erroneous. I am fully convinced that species are not immutable; but that those belonging to what are called the same genera are lineal descendants of some other and generally extinct species, in the same manner as the acknowledged varieties of any one species are the descendants of that species. Furthermore, I am convinced that Natural Selection has been the main but not exclusive means of modification.

Darwin opens his case for organic evolution with an argument by analogy from agriculture. Simply put, inherited variability is ubiquitous within species of domesticated animals and cultivated plants. The many varieties within the species arise in the breeder's program of propagating the most favored individuals, perhaps while culling out the others. Since selection in a breeding program operates mainly upon the most subtle variations, the tempo of the resulting change tends to be gradual. On its face, his use here of agriculture seems an inspired strategy; that this sword can cut both ways we shall eventually discover when we find it in the hands of Darwin's critics.

In Darwin's preliminary discussion of the sources of variability within species, note that the modern notion of genetic mutation upon which organic evolution rests is prefigured by his description of the "conditions of life" acting indirectly upon reproductive systems. He also gives some importance to the appearance of heritable traits acquired either directly as a consequence of environmental conditions or, in Lamarckian fashion, by their use and disuse. These latter two erroneous postulates, as we might expect with our unfailing hindsight, will lead Darwin into considerable difficulty in the defense of his theory.

CHAPTER I

Variation under Domestication.

WHEN WE LOOK to the individuals of the same variety or subvariety of our older cultivated plants and animals, one of the first points which strikes us, is, that they generally differ much more from each other, than do the individuals of any one species or variety in a state of nature. When we reflect on the vast diversity of the plants and animals which have been cultivated, and which have varied during all ages under the most different climates and treatment, I think we are driven to conclude that this greater variability is simply due to our domestic productions having been raised under conditions of life not so uniform as, and somewhat different from, those to which the parent-species have been exposed under nature. There is, also, I think, some probability in the view propounded by Andrew Knight, that this variability may be partly connected with excess of food. It seems pretty clear that organic beings must be exposed during several generations to the new conditions of life to cause any appreciable amount of variation; and that when the organisation has once begun to vary, it generally continues to vary for many generations. No case is on record of a variable being ceasing to be variable under cultivation. Our oldest cultivated plants, such as wheat, still often yield new varieties: our oldest domesticated animals are still capable of rapid improvement or modification.

Seedlings from the same fruit, and the young of the same litter, sometimes differ considerably from each other, though both the young and the parents, as Müller has remarked, have apparently been exposed to exactly the same conditions of life; and this shows how unimportant the direct effects of the conditions of life are in comparison with the laws of reproduction, and of growth, and of inheritance; for had the action of the conditions been direct, if any of the young had varied, all would probably have varied in the same manner. To judge how much, in the case of any variation, we should attribute to the direct action of heat, moisture, light, food, &c., is most difficult: my impression is, that with animals such agencies have produced very little direct effect, though apparently more in the case of plants. Under this point of view, Mr. Buckman's recent experiments on plants seem extremely valuable. When all or nearly all the individuals exposed to certain conditions are affected in the same way, the change at first appears to be directly due to such conditions; but in some cases it can be shown that quite opposite conditions produce similar changes of structure. Nevertheless some slight amount of change may, I think, be attributed to the direct action of the conditions of life — as, in some cases, increased size from amount of food, colour from particular kinds of food and from light, and perhaps the thickness of fur from climate.

Habit also has a decided influence, as in the period of flowering with plants when transported from one climate to another. In animals it has a more marked effect; for instance, I find in the domestic duck that the bones of the wing weigh less and the bones of the leg more, in proportion to the whole skeleton, than do the same bones in the wild-duck; and I presume that this change may be safely attributed to the domestic duck flying much less, and walking more, than its wild parent. The great and inherited development of the udders in cows and goats in countries where they are habitually milked, in comparison with the state of these organs in other countries, is another instance of the effect of use. Not a single domestic animal can be named which has not in some country drooping ears; and the view suggested by some authors, that the drooping is due to the disuse of the muscles of the ear, from the animals not being much alarmed by danger, seems probable.

There are many laws regulating variation, some few of which can be dimly seen, and will be hereafter briefly mentioned. I will here only allude to what may be called correlation of growth. Any change in the embryo or larva will almost certainly entail changes in the mature animal. . . . Hairless dogs have imperfect teeth; long-haired and coarse-haired animals are apt to have, as is asserted, long or many horns; pigeons with feathered feet have skin between their outer toes; pigeons with short beaks have small feet, and those with long beaks large feet. Hence, if a man goes on selecting, and thus augmenting, any peculiarity, he will almost certainly unconsciously modify other parts of the structure, owing to the mysterious laws of the correlation of growth.

Any variation which is not inherited is unimportant for us. But the number and diversity of inheritable deviations of structure, both those of slight and those of considerable physiological importance, is endless. Dr. Prosper Lucas's treatise, in two large volumes, is the fullest and the best on this subject. No breeder doubts how strong is the tendency to inheritance: like produces like is his fundamental belief: doubts have been thrown on this principle by theoretical writers alone. When a deviation appears not unfrequently, and we see it in the father and child, we cannot tell whether it may not be due to the same original cause acting on both; but when amongst individuals, apparently exposed to the same conditions, any very rare deviation, due to some extraordinary combination of circumstances, appears in the parent — say, once amongst several million individuals — and it reappears in the child, the mere doctrine of chances almost compels us to attribute its reappearance to inheritance. Every one must have heard of cases of albinism, prickly skin, hairy bodies, &c., appearing in several members of the same family. If strange and rare deviations of structure are truly inherited, less strange and commoner deviations may be freely admitted to be inheritable. Perhaps the correct way of viewing the whole subject, would be, to look at the inheritance of every character whatever as the rule, and non-inheritance as the anomaly.

Selection. — Let us now briefly consider the steps by which domestic races have been produced, either from one or from several allied species. Some little effect may, perhaps, be attributed to the direct action of the external conditions of life, and some little to habit; but he would be a bold man who would account by such agencies for the differences of a dray and race horse, a greyhound and bloodhound, a carrier and tumbler pigeon. One of the most remarkable features in our domesticated races is that we see in them adaptation, not indeed to the animal's or plant's own good, but to man's use or fancy. Some variations useful to him have probably arisen suddenly, or by one step; many botanists, for instance, believe that the fuller's teazle, with its hooks, which cannot be rivalled by any mechanical contrivance, is only a variety of the wild Dipsacus; and this amount of change may have suddenly arisen in a seedling. So it has probably been with the turnspit dog; and this is known to have been the case with the ancon sheep. But when we compare the dray-horse and race-horse, the dromedary and camel, the various breeds of sheep fitted either for cultivated land or mountain pasture, with the wool of one breed good for one purpose, and that of another breed for another purpose; when we compare the many breeds of dogs, each good for man in very different ways; when we compare the game-cock, so pertinacious in battle, with other breeds so little quarrelsome, with "everlasting layers" which never desire to sit, and with the bantam so small and elegant; when we compare the host of agricultural, culinary, orchard, and flower-garden races of plants, most useful to man at different seasons and for different purposes, or so beautiful in his eyes, we must, I think, look further than to mere variability. We cannot suppose that all the breeds were suddenly produced as perfect and as useful as we now see them; indeed, in several cases, we know that this has not been their history. The key is man's power of accumulative selection: nature gives successive variations; man adds them up in certain directions useful to him. In this sense he may be said to make for himself useful breeds.

What English breeders have actually effected is proved by the enormous prices given for animals with a good pedigree; and these have now been exported to almost every quarter of the world. The improvement is by no means generally due to crossing different breeds; all the best breeders are strongly opposed to this practice, except sometimes amongst closely allied sub-breeds. And when a cross has been made, the closest selection is far more indispensable even than in ordinary cases. If selection consisted merely in separating some very distinct variety, and breeding from it, the principle would be so obvious as hardly to be worth notice; but its importance consists in the great effect produced by the accumulation in one direction, during successive generations, of differences absolutely inappreciable by an uneducated eye — differences which I for one have vainly attempted to appreciate. Not one man in a thousand has accuracy of eye and judgment sufficient to become an eminent breeder. If gifted with these qualities, and he studies his subject for years, and devotes his lifetime to it with indomitable perseverance, he will succeed, and may make great improvements; if he wants any of these qualities, he will assuredly fail. Few would readily believe in the natural capacity and years of practice requisite to become even a skilful pigeon-fancier.

If there exist savages so barbarous as never to think of the inherited character of the offspring of their domestic animals, yet any one animal particularly useful to them, for any special purpose, would be carefully preserved during famines and other accidents, to which savages are so liable, and such choice animals would thus generally leave more offspring than the inferior ones; so that in this case there would be a kind of unconscious selection going on. We see the value set on animals even by the barbarians of Tierra del Fuego, by their killing and devouring their old women, in times of dearth, as of less value than their dogs.

In plants the same gradual process of improvement, through the occasional preservation of the best individuals, whether or not sufficiently distinct to be ranked at their first appearance as distinct varieties, and whether or not two or more species or races have become blended together by crossing, may plainly be recognised in the increased size and beauty which we now see in the varieties of the heartsease, rose, pelargonium, dahlia, and other plants, when compared with the older varieties or with their parent-stocks. No one would ever expect to get a first-rate heartsease or dahlia from the seed of a wild plant. No one would expect to raise a first-rate melting pear from the seed of the wild pear, though he might succeed from a poor seedling growing wild, if it had come from a garden-stock. The pear, though cultivated in classical times, appears, from Pliny's description, to have been a fruit of very inferior quality. I have seen great surprise expressed in horticultural works at the wonderful skill of gardeners, in having produced such splendid results from such poor materials; but the art, I cannot doubt, has been simple, and, as far as the final result is concerned, has been followed almost unconsciously. It has consisted in always cultivating the best known variety, sowing its seeds, and, when a slightly better variety has chanced to appear, selecting it, and so onwards. But the gardeners of the classical period, who cultivated the best pear they could procure, never thought what splendid fruit we should eat; though we owe our excellent fruit, in some small degree, to their having naturally chosen and preserved the best varieties they could anywhere find.

I must now say a few words on the circumstances, favourable, or the reverse, to man's power of selection. A high degree of variability is obviously favourable, as freely giving the materials for selection to work on; not that mere individual differences are not amply sufficient, with extreme care, to allow of the accumulation of a large amount of modification in almost any desired direction. But as variations manifestly useful or pleasing to man appear only occasionally, the chance of their appearance will be much increased by a large number of individuals being kept; and hence this comes to be of the highest importance to success. On this principle Marshall has remarked, with respect to the sheep of parts of Yorkshire, that "as they generally belong to poor people, and are mostly *in small lots*, they never can be improved." On the other hand, nurserymen, from raising large stocks of the same plants, are generally far more successful than amateurs in getting new and valuable varieties. The keeping of a large number of individuals of a species in any country requires that the species should be placed under favourable conditions of life, so as to breed freely in that country. When the individuals of any species are scanty, all the individuals, whatever their quality may be, will generally be allowed to breed, and this will effectually prevent selection. But probably the most important point of all, is, that the animal or plant should be so highly useful to man, or so much valued by him, that the closest attention should be paid to even the slightest deviation in the qualities or structure of each individual. Unless such attention be paid nothing can be effected.

To sum up on the origin of our Domestic Races of animals and plants. I believe that the conditions of life, from their action on the reproductive system, are so far of the highest importance as causing variability. I do not believe that variability is an inherent and necessary contingency, under all circumstances, with all organic beings, as some authors have thought. The effects of variability are modified by various degrees of inheritance and of reversion. Variability is governed by many unknown laws, more especially by that of correlation of growth. Something may be attributed to the direct action of the conditions of life. Something must be attributed to use and disuse. The final result is thus rendered infinitely complex. In some cases, I do not doubt that the intercrossing of species, aboriginally distinct, has played an important part in the origin of our domestic productions. When in any country several domestic breeds have once been established, their occasional intercrossing, with the aid of selection, has, no doubt, largely aided in the formation of new sub-breeds; but the importance of the crossing of varieties has, I believe, been greatly exaggerated, both in regard to animals and to those which are propagated by seed. In plants which are temporarily propagated by cuttings, buds, &c., the importance of the crossing both of distinct species and of varieties is immense; for the cultivator here quite disregards the extreme variability both of hybrids and mongrels, and the frequent sterility of hybrids; but the cases of plants not propagated by seed are of little importance to us, for their endurance is only temporary. Over all these causes of Change I am convinced that the accumulative action of Selection, whether applied methodically and more quickly, or unconsciously and more slowly, but more efficiently, is by far the predominant Power.

In this chapter Darwin begins to submit his circumstantial evidence for evolution in nature. Finding inherited variability to be as ubiquitous within natural species as within their domesticated counterparts, his particular concern here is to establish two points about the distribution of variability in nature that call out for an evolutionary interpretation. The first is simply that this variability tends to be expressed more or less continuously so as to present an "insensible gradation" between species as well as between their constituent subspecies or varieties. By blurring the distinction between the two levels, he seeks to convey an impression of a seamless evolutionary transition from variety to species: "species are only strongly marked and permanent varieties." There is a price to be paid in this argument, however, for it leads Darwin to discount the single feature that many of his contemporaries held — as we still do today — to define a living species: reproductive isolation, tending to restrict interbreeding to the confines of species boundaries.

The second point is that circumstances which favor the differentiation of separate varieties within species similarly promote the proliferation of separate species within genera. Here in support Darwin observes a tendency for genera with greater numbers of constituent species to contain species themselves more often subdivided by distinctly marked varieties. This point is really an extension of the first: varieties are incipient species and genera containing many species are likely to arise from species differentiated into many varieties. That is, a single mechanism acts to differentiate varieties into species and species into genera. Al-

though Darwin does not reveal until the next chapter what this differentiating mechanism might be, he does observe that the most geographically dispersed species tend also to be those containing the greatest number of varieties. We shall see in Chapter IV what role geographic dispersal plays in his theory.

CHAPTER II

Variation under Nature

BEFORE applying the principles arrived at in the last chapter to organic beings in a state of nature, we must briefly discuss whether these latter are subject to any variation. To treat this subject at all properly, a long catalogue of dry facts should be given; but these I shall reserve for my future work. Nor shall I here discuss the various definitions which have been given of the term species. No one definition has as yet satisfied all naturalists; yet every naturalist knows vaguely what he means when he speaks of a species. Generally the term includes the unknown element of a distinct act of creation. The term "variety" is almost equally difficult to define; but here community of descent is almost universally implied, though it can rarely be proved. We have also what are called monstrosities; but they graduate into varieties. By a monstrosity I presume is meant some considerable deviation of structure in one part, either injurious to or not useful to the species, and not generally propagated. Some authors use the term "variation" in a technical sense, as implying a modification directly due to the physical conditions of life; and "variations" in this sense are supposed not to be inherited: but who can say that the dwarfed condition of shells in the brackish waters of the Baltic, or dwarfed plants on Alpine summits, or the thicker fur of an animal from far northwards, would not in some cases be inherited for at least some few generations? and in this case I presume that the form would be called a variety.

Again, we have many slight differences which may be called individual differences, such as are known frequently to appear in the offspring from the same parents, or which may be presumed to have thus arisen, from being frequently observed in the individuals of the same species inhabiting the same confined locality. No one supposes that all the individuals of the same species are cast in the very same mould. These individual differences are highly important for us, as they afford materials for natural selection to accumulate, in the same manner as man can accumulate in any given direction individual differences in his domesticated productions. These individual differences generally affect what naturalists consider unimportant parts; but I could show by a long catalogue of facts, that parts which must be called important, whether viewed under a physiological or classificatory point of view, sometimes vary in the individuals of the same species. I am convinced that the most experienced naturalist would be surprised at the number of the cases of variability, even in important parts of structure, which he could collect on good authority, as I have collected, during a course of years. It should be remembered that systematists are far from pleased at finding variability in important characters, and that there are not many men who will laboriously examine internal and important organs, and compare them in many specimens of the same species. I should never have expected that the branching of the main nerves close to the great central ganglion of an insect would have been variable in the same species; I should have expected that changes of this nature could have been effected only by slow degrees: yet quite recently Mr. Lubbock has shown a degree of variability in these main nerves in Coccus, which may almost be compared to the irregular branching of the stem of a tree. This philosophical naturalist, I may add, has also quite recently shown that the muscles in the larvae of certain insects are very far from uniform. Authors sometimes argue in a circle when they state that important organs never vary; for these same authors practically rank that character as important (as some few naturalists have honestly confessed) which does not vary; and, under this point of view, no instance of an important part varying will ever be found: but under any other point of view many instances assuredly can be given.

*

Those forms which possess in some considerable degree the character of species, but which are so closely similar to some other forms, or are so closely linked to them by intermediate gradations, that naturalists do not like to rank them as distinct species, are in several respects the most important for us. We have every reason to believe that many of these doubtful and closely-allied forms have permanently retained their characters in their own country for a long time; for as long, as far as we know, as have good and true species. Practically, when a naturalist can unite two forms together by others having intermediate characters, he treats the one as a variety of the other, ranking the most common, but sometimes the one first described, as the species, and the other as the variety. But cases of great difficulty, which I will not here enumerate, sometimes occur in deciding whether or not to rank one form as a variety of another, even when they are closely connected by intermediate links; nor will the commonly-assumed hybrid nature of the intermediate links always remove the difficulty. In very many cases, however, one form is ranked as a variety of another, not because the intermediate links have actually been found, but because analogy leads the observer to suppose either that they do now somewhere exist, or may formerly have existed; and here a wide door for the entry of doubt and conjecture is opened.

*

Certainly no clear line of demarcation has as yet been drawn between species and sub-species — that is, the forms which in the opinion of some naturalists come very near to, but do not quite arrive at the rank of species; or, again, between sub-species and well-marked varieties, or between lesser varieties and individual differences. These differences blend into each other in an insensible series; and a series impresses the mind with the idea of an actual passage.

Hence I look at individual differences, though of small interest to the systematist, as of high importance for us, as being the first step towards such slight varieties as are barely thought worth recording in works on natural history. And I look at varieties which are in any degree more distinct and permanent, as steps leading to more strongly marked and more permanent varieties; and at these latter, as leading to sub-species, and to species. The passage from one stage of difference to another and higher stage may be, in some cases, due merely to the long-continued action of different physical conditions in two different regions; but I have not much faith in this view; and I attribute the passage of a variety, from a state in which it differs very slightly from its parent to one in which it differs more, to the action of natural selection in accumulating (as will hereafter be more fully explained) differences of structure in certain definite directions. Hence I believe a well-marked variety may be justly called an incipient species; but whether this belief be justifiable must be judged of by the general weight of the several facts and views given throughout this work.

It need not be supposed that all varieties or incipient species necessarily attain the rank of species. They may whilst in this incipient state become extinct, or they may endure as varieties for very long periods, as has been shown to be the case by Mr. Wollaston with the varieties of certain fossil land-shells in Madeira. If a variety were to flourish so as to exceed in numbers the parent species, it would then rank as the species, and the species as the variety; or it might come to supplant and exterminate the parent species; or both might co-exist, and both rank as independent species. But we shall hereafter have to return to this subject.

From these remarks it will be seen that I look at the term species, as one arbitrarily given for the sake of convenience to a set of individuals closely resembling each other, and that it does not essentially differ from the term variety, which is given to less distinct and more fluctuating forms. The term variety, again, in comparison with mere individual differences, is also applied arbitrarily, and for mere convenience sake.

From looking at species as only strongly-marked and well-defined varieties, I was led to anticipate that the species of the larger genera in each country would oftener present varieties, than the species of the smaller genera; for wherever many closely related species (*i.e.* species of the same genus) have been formed, many varieties or incipient species ought, as a general rule, to be now forming. Where many large trees grow, we expect to find saplings. Where many species of a genus have been formed through variation, circumstances have been favourable for variation; and hence we might expect that the circumstances would generally be still favourable to variation. On the other hand, if we look at each species as a special act of creation, there is no apparent reason why more varieties should occur in a group having many species, than in one having few.

To test the truth of this anticipation I have arranged the plants of twelve countries, and the coleopterous insects of two districts, into two nearly equal masses, the species of the larger genera on one side, and those of the smaller genera on the other side, and it has invariably proved to be the case that a larger proportion of the species on the side of the larger genera present varieties, than on the side of the smaller genera. Moreover, the species of the large genera which present any varieties, invariably present a larger average number of varieties than do the species of the small genera. Both these results follow when another division is made, and when all the smallest genera, with from only one to four species, are absolutely excluded from the tables. These facts are of plain signification on the view that species are only strongly marked and permanent varieties; for wherever many species of the same genus have been formed, or where, if we may use the expression, the manufactory of species has been active, we ought generally to find the manufactory still in action, more especially as we have every reason to believe the process of manufacturing new species to be a slow one. And this certainly is the case, if varieties be looked at as incipient species; for my tables clearly show as a general rule that, wherever many species of a genus have been formed, the species of that genus present a number of varieties, that is of incipient species, beyond the average. It is not that all large genera are now varying much, and are thus increasing in the number of their species, or that no small genera are now varying and increasing; for if this had been so, it would have been fatal to my theory; inasmuch as geology plainly tells us that small genera have in the lapse of time often increased greatly in size; and that large genera have often come to their maxima, declined, and disappeared. All that we want to show is, that where many species of a genus have been formed, on an average many are still forming; and this holds good.

There are other relations between the species of large genera and their recorded varieties which deserve notice. We have seen that there is no infallible criterion by which to distinguish species and well-marked varieties; and in those cases in which intermediate links have not been found between doubtful forms, naturalists are compelled to come to a determination by the amount of difference between them, judging by analogy whether or not the amount suffices to raise one or both to the rank of species. Hence the amount of difference is one very important criterion in settling whether two forms should be ranked as species or varieties. Now Fries has remarked in regard to plants, and Westwood in regard to insects, that in large genera the amount of difference between the species is often exceedingly small. I have endeavoured to test this numerically by averages, and, as far as my imperfect results go, they always confirm the view. I have also consulted some sagacious and most experienced observers, and, after deliberation, they concur in this view. In this respect, therefore, the species of the larger genera resemble varieties, more than do the species of the smaller genera. Or the case may be put in another way, and it may be said, that in the larger genera, in which a number of varieties or incipient species greater than the average are now manufacturing, many of the species already manufactured still to a certain extent resemble varieties, for they differ from each other by a less than usual amount of difference.

Moreover, the species of the large genera are related to each other, in the same manner as the varieties of any one species are related to each other. No naturalist pretends that all the species of a genus are equally distinct from each other; they may generally be divided into sub-genera, or sections, or lesser groups. As Fries has well remarked, little groups of species are generally clustered like satellites around certain other species. And what are varieties but groups of forms, unequally related to each other, and clustered round certain forms — that is, round their parent-species? Undoubtedly there is one most important point of difference between varieties and species; namely that the amount of difference between varieties, when compared with each other or with their parent-species, is much less than that between the species of the same genus. But when we come to discuss the principle, as I call it, of Divergence of Character, we shall see how this may be explained, and how the lesser differences between varieties will tend to increase into the greater differences between species.

Finally, then, varieties have the same general characters as species, for they cannot be distinguished from species, — except, firstly, by the discovery of intermediate linking forms, and the occurrence of such links cannot affect the actual characters of the forms which they connect; and except, secondly, by a certain amount of difference, for two forms, if differing very little, are generally ranked as varieties, notwithstanding that intermediate linking forms have not been discovered; but the amount of difference considered necessary to give to two forms the rank of species is quite indefinite. In genera having more than the average number of species in any country, the species of these genera have more than the average number of varieties. In large genera the species are apt to be closely, but unequally, allied together, forming little clusters round certain species. Species very closely allied to other species apparently have restricted ranges. In all these several respects the species of large genera present a strong analogy with varieties. And we can clearly understand these analogies, if species have once existed as varieties, and have thus originated: whereas, these analogies are utterly inexplicable if each species has been independently created.

We have, also, seen that it is the most flourishing and dominant species of the larger genera which on an average vary most; and varieties, as we shall hereafter see, tend to become converted into new and distinct species. The larger genera thus tend to become larger; and throughout nature the forms of life which are now dominant tend to become still more dominant by leaving many modified and dominant descendants. But by steps hereafter to be explained, the larger genera also tend to break up into smaller genera. And thus, the forms of life throughout the universe become divided into groups subordinate to groups.

In this chapter and in the next emerges the core element in Darwin's theory — Natural Selection. Here Darwin sets its stage, describing a relentless struggle for existence among species, a struggle flowing from their improvidential tendency to expand in numbers beyond the capacity of the natural economy. The reader interested in the importance of this Malthusian thesis to Darwin's formulation would benefit from the discussion in Ospovat.[1] Note that while Darwin emphasizes the most salient character of this struggle — the sometimes ferocious competition among individuals for survival — he is quick to translate its outcome into the success of its players in contributing offspring to the succeeding generation. Reproductive success is thus the keystone of natural selection. Along with the final one, this chapter contains some of the Origin's *most lyrical prose; there is a special poignancy in his concluding paragraph, reassuring us that nature's strife is all for the best.*

CHAPTER III

Struggle for Existence

BEFORE entering on the subject of this chapter, I must make a few preliminary remarks, to show how the struggle for existence bears on Natural Selection. It has been seen in the last chapter that amongst organic beings in a state of nature there is some individual variability; indeed I am not aware that this has ever been disputed. It is immaterial for us whether a multitude of doubtful forms be called species or sub-species or varieties; what rank, for instance, the two or three hundred doubtful forms of British plants are entitled to hold, if the existence of any well-marked varieties be admitted. But the mere existence of individual variability and of some few well-marked varieties, though necessary as the foundation for the work, helps us but little in understanding how species arise in nature. How have all those exquisite adaptations of one part of the organisation to another part, and to the conditions of life, and of one distinct organic being to another being, been perfected? We see these beautiful co-adaptations most plainly in the woodpecker and missletoe; and only a little less plainly in the humblest parasite which clings to the hairs of a quadruped or feathers of a bird; in the structure of the beetle which dives through the water; in the plumed seed which is wafted by the gentlest breeze; in short, we see beautiful adaptations everywhere and in every part of the organic world.

Again, it may be asked, how is it that varieties, which I have called incipient species, become ultimately converted into good and distinct species, which in most cases obviously differ from each other far more than do the varieties of the same species? How do those groups of species, which constitute what are called distinct genera, and which differ from each other more than do the species of the same genus, arise? All these results, as we shall more fully see in the next chapter, follow inevitably from the struggle for life. Owing to this struggle for life, any variation, however slight and from whatever cause proceeding, if it be in any degree profitable to an individual of any species, in its infinitely complex relations to other organic beings and to external nature, will tend to the preservation of that individual, and will generally be inherited by its offspring. The offspring, also, will thus have a better chance of surviving, for, of the many individuals of any species which are periodically born, but a small number can survive. I have called this principle, by which each slight variation, if useful, is preserved, by the term of Natural Selection, in order to mark its relation to man's power of selection. We have seen that man by selection can certainly produce great results, and can adapt organic beings to his own uses, through the accumulation of slight but useful variations, given to him by the hand of Nature. But Natural

Selection, as we shall hereafter see, is a power incessantly ready for action, and is as immeasurably superior to man's feeble efforts, as the works of Nature are to those of Art.

We will now discuss in a little more detail the struggle for existence. In my future work this subject shall be treated, as it well deserves, at much greater length. The elder De Candolle and Lyell have largely and philosophically shown that all organic beings are exposed to severe competition. In regard to plants, no one has treated this subject with more spirit and ability than W. Herbert, Dean of Manchester, evidently the result of his great horticultural knowledge. Nothing is easier than to admit in words the truth of the universal struggle for life, or more difficult — at least I have found it so — than constantly to bear this conclusion in mind. Yet unless it be thoroughly engrained in the mind, I am convinced that the whole economy of nature, with every fact on distribution, rarity, abundance, extinction, and variation, will be dimly seen or quite misunderstood. We behold the face of nature bright with gladness, we often see superabundance of food; we do not see, or we forget, that the birds which are idly singing round us mostly live on insects or seeds, and are thus constantly destroying life; or we forget how largely these songsters, or their eggs, or their nestlings, are destroyed by birds and beasts of prey; we do not always bear in mind, that though food may be now superabundant, it is not so at all seasons of each recurring year.

I should premise that I use the term Struggle for Existence in a large and metaphorical sense, including dependence of one being on another, and including (which is more important) not only the life of the individual, but success in leaving progeny. Two canine animals in a time of dearth, may be truly said to struggle with each other which shall get food and live. But a plant on the edge of a desert is said to struggle for life against the drought, though more properly it should be said to be dependent on the moisture. A plant which annually produces a thousand seeds, of which on an average only one comes to maturity, may be more truly said to struggle with the plants of the same and other kinds which already clothe the ground. The missletoe is dependent on the apple and a few other trees, but can only in a far-fetched sense be said to struggle with these trees, for if too many of these parasites grow on the same tree, it will languish and die. But several seedling missletoes, growing close together on the same branch, may more truly be said to struggle with each other. As the missletoe is disseminated by birds, its existence depends on birds; and it may metaphorically be said to struggle with other fruit-bearing plants, in order to tempt birds to devour and thus disseminate its seeds rather than those of other plants. In these several senses, which pass into each other, I use for convenience sake the general term of struggle for existence.

A struggle for existence inevitably follows from the high rate at which all organic beings tend to increase. Every being, which during its natural lifetime produces several eggs or seeds, must suffer destruction during some period of its life, and during some season or occasional year, otherwise, on the principle of geometrical increase, its numbers would quickly become so inordinately great that no country could support the product. Hence, as more individuals are produced than can possibly survive, there must in every case be a struggle for existence, either one individual with another of the same species, or with the individuals of distinct species, or with the physical conditions of life. It is the doctrine of Malthus applied with manifold force to the whole animal and vegetable kingdoms; for in this case there can be no artificial increase of food, and no prudential restraint from marriage. Although some species may be now increasing, more or less rapidly, in numbers, all cannot do so, for the world would not hold them.

There is no exception to the rule that every organic being naturally increases at so high a rate, that if not destroyed, the earth would soon be covered by the progeny of a single pair. Even slow-breeding man has doubled in twenty-five years, and at this rate, in a few thousand years, there would literally not be standing room for his progeny. Linnaeus has calculated that if an annual plant produced only two seeds — and there is no plant so unproductive as this — and their seedlings next year produced two, and so on, then in twenty years there would be a million plants. The elephant is reckoned to be the slowest breeder of all known animals, and I have taken some pains to estimate its probable minimum rate of natural increase: it will be under the mark to assume that it breeds when thirty years old, and goes on breeding till ninety years old, bringing forth three pair of young in this interval; if this be so, at the end of the fifth century there would be alive fifteen million elephants, descended from the first pair.

In a state of nature almost every plant produces seed, and amongst animals there are very few which do not annually pair. Hence we may confidently assert, that all plants and animals are tending to increase at a geometrical ratio, that all would most rapidly stock every station in which they could any how exist, and that the geometrical tendency to increase must be checked by destruction at some period of life. Our familiarity with the larger domestic animals tends, I think, to mislead us: we see no great destruction falling on them, and we forget that thousands are annually slaughtered for food, and that in a state of nature an equal number would have somehow to be disposed of.

The only difference between organisms which annually produce eggs or seeds by the thousand, and those which produce extremely few, is that the slow-breeders would require a few more years to people, under favourable conditions, a whole district, let it be ever so large. The condor lays a couple of eggs and the ostrich a score, and yet in the same country the condor may be the more numerous of the two: the Fulmar petrel lays but one egg, yet it is believed to be the most numerous bird in the world. One fly deposits hundreds of eggs, and another, like the hippobosca, a single one; but this difference does not determine how many individuals of the two species can be supported in a district. A large number of eggs is of some importance to those species, which depend on a rapidly fluctuating amount of food, for it allows them rapidly to increase in number. But the real importance of a large number of eggs or seeds is to make up for much destruction at some period of life; and this period in the great majority of cases is an early one. If an animal can in any way protect its own eggs or young, a small number may be produced, and yet the average stock be fully kept up; but if many eggs or young are destroyed, many must be produced, or the species will become extinct. It would suffice to keep up the full number of a tree, which lived on an average for a thousand years, if a single seed were produced once in a thousand years, supposing that this seed were never destroyed, and could be ensured to germinate in a fitting place. So that in all cases, the average number of any animal or plant depends only indirectly on the number of its eggs or seeds.

In looking at Nature, it is most necessary to keep the foregoing considerations always in mind — never to forget that every single organic being around us may be said to be striving to the utmost to increase in numbers; that each lies by a struggle at some period of its life; that heavy destruction inevitably falls either on the young or old, during each generation or at recurrent intervals. Lighten any check, mitigate the destruction ever so little, and the number of the species will almost instantaneously increase to any amount. The face of Nature may be compared to a yielding surface, with ten thousand sharp wedges packed close together and driven inwards by incessant blows, sometimes one wedge being struck, and then another with greater force.

What checks the natural tendency of each species to increase in number is most obscure. Look at the most vigorous species; by as much as it swarms in numbers, by so much will its tendency to increase be still further increased. We know not exactly what the checks are in even one single instance. Nor will this surprise any one who reflects how ignorant we are on this head, even in regard to mankind, so incomparably better known than any other animal.

The amount of food for each species of course gives the extreme limit to which each can increase; but very frequently it is not the obtaining food, but the serving as prey to other animals, which determines the average numbers of a species. Thus, there seems to be little doubt that the stock of partridges, grouse, and hares on any large estate depends chiefly on the destruction of vermin. If not one head of game were shot during the next twenty years in England, and, at the same time, if no vermin were destroyed, there would, in all probability, be less game than at present, although hundreds of thousands of game animals are now annually killed. On the other hand, in some cases, as with the elephant and rhinoceros, none are destroyed by beasts of prey: even the tiger in India most rarely dares to attack a young elephant protected by its dam.

Climate plays an important part in determining the average numbers of a species, and periodical seasons of extreme cold or drought, I believe to be the most effective of all checks. I estimated that the winter of 1854–55 destroyed four-fifths of the birds in my own grounds; and this is a tremendous destruction, when we remember that ten per cent. is an extraordinarily severe mortality from epidemics with man. The action of climate seems at first sight to be quite independent of the struggle for existence; but in so far as climate chiefly acts in reducing food, it brings on the most severe struggle between the individuals, whether of the same or of distinct species, which subsist on the same kind of food. Even when climate, for instance extreme cold, acts directly, it will be the least vigorous, or those which have got least food through the advancing winter, which will suffer most. When we travel from south to north, or from a damp region to a dry, we invariably see some species gradually getting rarer and rarer, and finally disappearing; and the change of climate being conspicuous, we are tempted to attribute the whole effect to its direct action. But this is a very false view: we forget that each species, even where it most abounds, is constantly suffering enormous destruction at some period of its life, from enemies or from competitors for the same place and food; and if these enemies or competitors be in the least degree favoured by any slight change of climate, they will increase in numbers, and, as each area is already fully stocked with inhabitants, the other species will decrease. When we travel southward and see a species decreasing in numbers, we may feel sure that the cause lies quite as much in other species being favoured, as in this one being hurt. So it is when we travel northward, but in a somewhat lesser degree, for the number of species of all kinds, and therefore of competitors, decreases northwards; hence in going northward, or in ascending a mountain, we far oftener meet with stunted forms, due to the *directly* injurious action of climate, than we do in proceeding southwards or in descending a mountain. When we reach the Arctic regions, or snow-capped summits, or absolute deserts, the struggle for life is almost exclusively with the elements.

That climate acts in main part indirectly by favouring other species, we may clearly see in the prodigious number of plants in our gardens which can perfectly well endure our climate, but which never become naturalised, for they cannot compete with our native plants, nor resist destruction by our native animals.

When a species, owing to highly favourable circumstances, increases inordinately in numbers in a small tract, epidemics — at least, this seems generally to occur with our game animals — often ensue: and here we have a limiting check independent of the struggle for life. But even some of these so-called epidemics appear to be due to parasitic worms, which have from some cause, possibly in part through facility of diffusion amongst the crowded animals, been disproportionably favoured: and here comes in a sort of struggle between the parasite and its prey.

Many cases are on record showing how complex and unexpected are the checks and relations between organic beings, which have to struggle together in the same country. I will give only a single instance, which, though a simple one, has interested me. In Staffordshire, on the estate of a relation where I had ample means of investigation, there was a large and extremely barren heath, which had never been touched by the hand of man; but several hundred acres of exactly the same nature had been enclosed twenty-five years previously and planted with Scotch fir. The change in the native vegetation of the planted part of the heath was most remarkable, more than is generally seen in passing from one quite different soil to another: not only the proportional numbers of the heath-plants were wholly changed, but twelve species of plants (not counting grasses and carices) flourished in the plantations, which could not be found on the heath. The effect on the insects must have been still greater, for six insectivorous birds were very common in the plantations, which were not to be seen on the heath; and the heath was frequented by two or three distinct insectivorous birds. Here we see how potent has been the effect of the introduction of a single tree, nothing whatever else having been done, with the exception that the land had been enclosed, so that cattle could not enter. But how important an element enclosure is, I plainly saw near Farnham, in Surrey. Here there are extensive heaths, with a few clumps of old Scotch firs on the distant hilltops: within the last ten years large spaces have been enclosed, and self-sown firs are now springing up in multitudes, so close together that all cannot live. When I ascertained that these young trees had not been sown or planted,

I was so much surprised at their numbers that I went to several points of view, whence I could examine hundreds of acres of the unenclosed heath, and literally I could not see a single Scotch fir, except the old planted clumps. But on looking closely between the stems of the heath, I found a multitude of seedlings and little trees, which had been perpetually browsed down by the cattle. In one square yard, at a point some hundred yards distant from one of the old clumps, I counted thirty-two little trees; and one of them, judging from the rings of growth, had during twenty-six years tried to raise its head above the stems of the heath, and had failed. No wonder that, as soon as the land was enclosed, it became thickly clothed with vigorously growing young firs. Yet the heath was so extremely barren and so extensive that no one would ever have imagined that cattle would have so closely and effectually searched it for food.

Here we see that cattle absolutely determine the existence of the Scotch fir; but in several parts of the world insects determine the existence of cattle. Perhaps Paraguay offers the most curious instance of this; for here neither cattle nor horses nor dogs have ever run wild, though they swarm southward and northward in a feral state; and Azara and Rengger have shown that this is caused by the greater number in Paraguay of a certain fly, which lays its eggs in the navels of these animals when first born. The increase of these flies, numerous as they are, must be habitually checked by some means, probably by birds. Hence, if certain insectivorous birds (whose numbers are probably regulated by hawks or beasts of prey) were to increase in Paraguay, the flies would decrease — then cattle and horses would become feral, and this would certainly greatly alter (as indeed I have observed in parts of South America) the vegetation: this again would largely affect the insects; and this, as we just have seen in Staffordshire, the insectivorous birds, and so onwards in ever-increasing circles of complexity. We began this series by insectivorous birds, and we have ended with them. Not that in nature the relations can ever be as simple as this.

Battle within battle must ever be recurring with varying success; and yet in the long-run the forces are so nicely balanced, that the face of nature remains uniform for long periods of time, though assuredly the merest trifle would often give the victory to one organic being over another. Nevertheless so profound is our ignorance, and so high our presumption, that we marvel when we hear of the extinction of an organic being; and as we do not see the cause, we invoke cataclysms to desolate the world, or invent laws on the duration of the forms of life!

In the case of every species, many different checks, acting at different periods of life, and during different seasons or years, probably come into play; some one check or some few being generally the most potent, but all concurring in determining the average number or even the existence of the species. In some cases it can be shown that widely-different checks act on the same species in different districts. When we look at the plants and bushes clothing an entangled bank, we are tempted to attribute their proportional numbers and kinds to what we call chance. But how false a view is this! Every one has heard that when an American forest is cut down, a very different vegetation springs up; but it has been observed that the trees now growing on the ancient Indian mounds, in the Southern United States, display the same beautiful diversity and proportion of kinds as in the surrounding virgin forests. What a struggle between the several kinds of trees must here have gone on during long centuries, each annually scattering its seeds by the thousand; what war between insect and insect — between insects, snails, and other animals with birds and beasts of prey — all striving to increase, and all feeding on each other or on the trees or their seeds and seedlings, or on the other plants which first clothed the ground and thus checked the growth of the trees! Throw up a handful of feathers, and all must fall to the ground according to definite laws; but how simple is this problem compared to the action and reaction of the innumerable plants and animals which have determined, in the course of centuries, the proportional numbers and kinds of trees now growing on the old Indian ruins!

The dependency of one organic being on another, as of a parasite on its prey, lies generally between beings remote in the scale of nature. This is often the case with those which may strictly be said to struggle with each other for existence, as in the case of locusts and grass-feeding quadrupeds. But the struggle almost invariably will be most severe between the individuals of the same species, for they frequent the same districts, require the same food, and are exposed to the same dangers. In the case of varieties of the same species, the struggle will generally be almost equally severe, and we sometimes see the contest soon decided: for instance, if several varieties of wheat be sown together, and the mixed seed be resown, some of the varieties which best suit the soil or climate, or are naturally the most fertile, will beat the others and so yield more seed, and will consequently in a few years quite supplant the other varieties.

As species of the same genus have usually, though by no means invariably, some similarity in habits and constitution, and always in structure, the struggle will generally be more severe between species of the same genus, when they come into competition with each other, than between species of distinct genera. We see this in the recent extension over parts of the United States of one species of swallow having caused the decrease of another species. The recent increase of the misselthrush in parts of Scotland has caused the decrease of the songthrush. How frequently we hear of one species of rat taking the place of another species under the most different climates! In Russia the small Asiatic cockroach has everywhere driven before it its great congener. One species of charlock will supplant another, and so in other cases. We can dimly see why the competition should be most severe between allied forms, which fill nearly the same place in the economy of nature; but probably in no one case could we precisely say why one species has been victorious over another in the great battle of life.

A corollary of the highest importance may be deduced from the foregoing remarks, namely, that the structure of every organic being is related, in the most essential yet often hidden manner, to that of all other organic beings, with which it comes into competition for food or residence, or from which it has to escape, or on which it preys. This is obvious in the structure of the teeth and talons of the tiger; and in that of the legs and claws of the parasite which clings to the hair on the tiger's body. But in the beautifully plumed seed of the dandelion, and in the flattened and fringed legs of the water-beetle, the relation seems at first confined to the elements of air and water. Yet the advantage of plumed seeds no doubt stands in the closest relation to the land being already thickly clothed by other plants; so that the seeds may be widely distributed and fall on unoccupied ground. In the water-beetle, the structure of its legs, so well adapted for diving, allows it to compete with other aquatic insects, to hunt for its own prey, and to escape serving as prey to other animals.

The store of nutriment laid up within the seeds of many plants seems at first sight to have no sort of relation to other plants. But from the strong growth of young plants produced from such seeds (as peas and beans), when sown in the midst of long grass, I suspect that the chief use of the nutriment in the seed is to favour the growth of the young seedling, whilst struggling with other plants growing vigorously all around.

Hence, also, we can see that when a plant or animal is placed in a new country amongst new competitors, though the climate may be exactly the same as in its former home, yet the conditions of its life will generally be changed in an essential manner. If we wished to increase its average numbers in its new home, we should have to modify it in a different way to what we should have done in its native country; for we should have to give it some advantage over a different set of competitors or enemies.

It is good thus to try in our imagination to give any form some advantage over another. Probably in no single instance should we know what to do, so as to succeed. It will convince us of our ignorance on the mutual relations of all organic beings; a conviction as necessary, as it seems to be difficult to acquire. All that we can do, is to keep steadily in mind that each organic being is striving to increase at a geometrical ratio; that each at some period of its life, during some season of the year, during each generation or at interals, has to struggle for life, and to suffer great destruction. When we reflect on this struggle, we may console ourselves with the full belief, that the war of nature is not incessant, that no fear is felt, that death is generally prompt, and that the vigorous, the healthy, and the happy survive and multiply.

Coupling inherited diversity to the struggle for survival, at last Darwin gives us Natural Selection. Since those individuals whose inherited character confers upon them some slight advantage in their struggle for survival are also those more likely to leave offspring in greater number, the cumulative outcome is a gradual and continual shift within the species toward an increasing degree of adaptation to the environment. Natural selection thus assumes the creative governance of the evolutionary process. The form of this argument is quintessentially Lyellian, translated from geological to biological processes (see page 102): observable, small-

scale phenomena act "silently and insensibly" over vast periods to produce monumental change.

Again with an eye to examining the reception subsequently given the Origin by its critics, several additional points Darwin raises here are worth remarking. The first is that while he recognizes that geographic isolation might in some instances favor the differentiation of new species (by checking the influx of immigrants), on balance he finds large, widely dispersed populations more advantageous yet to the formation of new species. Since this runs counter to modern evolutionary theory, with its emphasis upon reproductive isolation in the formation of new species by splitting, it is instructive to note what Darwin's reasons were. The most obvious is that the variations upon which he held natural selection to operate are more likely to arise within larger populations. The less evident reason stems from his view that species dispersed broadly are subject to more intense competitive pressures from other species. Darwin believed that there would be a greater diversity of habitats — and species already evolved to occupy them — over a wide and unconfined geographic range. Consequently, evolution ought to proceed more rapidly under these conditions. But Darwin was wrong on both scores, and even those friendly to his evolutionary thesis could not countenance his dismissal of geographic isolation as unnecessary to the founding of new species.

A second point, and here we shall side with Darwin, is that competition leads to the episodic extinction of groups that are less well adapted, and where the competition applies between an ancestral species and its better-adapted evolutionary descendants, the latter tend to endure. By this scheme, then, ancestral species are replaced by their descendant species. Thus, by the elimination of transitional forms, successive evolutionary lineages tend to diverge in their character. The cumulative effect, as Darwin describes in his metaphor of the Tree of Life, is to render ever greater the separation of its major branches. Never-

theless, several of his critics failed entirely to grasp this argument, complaining that the terms of his theory lead us to expect a continuum of living species linking, say, reptiles and birds. Yet it ought to be plain from Darwin's explanation that such early transitional forms that did exist have given way to their better-adapted descendants, and that such criticism is without justification.

Finally, despite the primacy he accords natural selection, Darwin introduces here, in addition to the Lamarckian principle of use and disuse, yet a third evolutionary mechanism, sexual selection. We shall have reason later to consider its details, particularly insofar as it competes with natural selection to explain evolution.

CHAPTER IV

Natural Selection

How will the struggle for existence, discussed too briefly in the last chapter, act in regard to variation? Can the principle of selection, which we have seen is so potent in the hands of man, apply in nature? I think we shall see that it can act most effectually. Let it be borne in mind in what an endless number of strange peculiarities our domestic productions and, in a lesser degree, those under nature, vary; and how strong the hereditary tendency is. Under domestication, it may be truly said that the whole organisation becomes in some degree plastic. Let it be borne in mind how infinitely complex and close-fitting are the mutual relations of all organic beings to each other and to their physical conditions of life. Can it, then, be thought improbable, seeing that variations useful to man have undoubtedly occurred, that other variations useful in some way to each being in the great and complex battle of life should sometimes occur in the course of thousands of generations? If such do occur, can we doubt (remembering that many more individuals are born than can possibly survive) that individuals having any advantage, however slight, over others, would have the best chance of surviving and of procreating their kind? On the other hand, we may feel sure that any variation in the least degree injurious would be rigidly destroyed. This preservation of favourable variations and the rejection of injurious variations, I call Natural Selection. Variations neither useful nor injurious would not be affected by natural selection, and would be left a fluctuating element, as perhaps we see in the species called polymorphic.

We shall best understand the probable course of natural selection by taking the case of a country undergoing some physical change, for instance, of climate. The proportional numbers of its inhabitants would almost immediately undergo a change, and some species might become extinct. We may conclude, from what we have seen of the intimate and complex manner in which the inhabitants of each country are bound together, that any change in the numerical proportions of some of the inhabitants, independently of the change of climate itself, would most seriously affect many of the others. If the country were open on its borders, new forms would certainly immigrate, and this also would seriously disturb the relations of some of the former inhabitants. Let it be remembered how powerful the influence of a single introduced tree or mammal has been shown to be. But in the case of an island, or of a country partly surrounded by barriers, into which new and better adapted forms could not freely enter, we should then have places in the economy of nature which would assuredly be better filled up, if some of the original inhabitants were in some manner modified; for, had the area been open to immigration, these same places would have been seized on by intruders. In such case, every slight modification, which in the course of ages chanced to arise, and which in any way favoured the individuals of any of the species, by better adapting them to their altered conditions, would tend to be preserved; and natural selection would thus have free scope for the work of improvement.

We have reason to believe, as stated in the first chapter, that a change in the conditions of life, by specially acting on the reproductive system, causes or increases variability; and in the foregoing case the conditions of life are supposed to have undergone a change, and this would manifestly be favourable to natural selection, by giving a better chance of profitable variations occurring; and unless profitable variations do occur, natural selection can do nothing. Not that, as I believe, any extreme amount of variability is necessary; as man can certainly produce great results by adding up in any given direction mere individual differences, so could Nature, but far more easily, from having incomparably longer time at her disposal. Nor do I believe that any great physical change, as of climate, or any unusual degree of isolation to check immigration, is actually necessary to produce new and unoccupied places for natural selection to fill up by modifying and improving some of the varying inhabitants. For as all the inhabitants of each country are struggling together with nicely balanced forces, extremely slight modifications in the structure or habits of one inhabitant would often give it an advantage over others; and still further modifications of the same kind would often still further increase the advantage. No country can be named in which all the native inhabitants are now so perfectly adapted to each other and to the physical conditions under which they live, that none of them could anyhow be improved; for in all countries, the natives have been so far conquered by naturalised productions, that they have allowed foreigners to take firm possession of the land. And as foreigners have thus everywhere beaten some of the natives, we may safely conclude that the natives might have been modified with advantage, so as to have better resisted such intruders.

As man can produce and certainly has produced a great result by his methodical and unconscious means of selection, what may not nature effect? Man can act only on external and visible characters: nature cares nothing for appearances, except in so far as they may be useful to any being. She can act on every internal organ, on every shade of constitutional difference, on the whole machinery of life. Man selects only for his own good; Nature only for that of the being which she tends. Every selected character is fully exercised by her; and the being is placed under well-suited conditions of life. Man keeps the natives of many climates in the same country; he seldom exercises each selected character in some peculiar and fitting manner; he feeds a long and a short beaked pigeon on the same food; he does not exercise a long-backed or long-legged quadruped in any peculiar manner; he exposes sheep with long and short wool to the same climate. He does not allow the most vigorous males to struggle for the females. He does not rigidly destroy all inferior animals, but protects during each varying season, as far as lies in his power, all his productions. He often begins his selection by some half-monstrous form; or at least by some modification prominent enough to catch his eye, or to be plainly useful to him. Under nature, the slightest difference of structure or constitution may well turn the nicely-balanced scale in the struggle for life, and so be preserved. How fleeting are the wishes and efforts of man! how short his time! and consequently how poor will his products be, compared with those accummulated by nature during whole geological periods. Can we wonder, then, that nature's productions should be far "truer" in character than man's productions; that they should be infinitely better adapted to the most complex conditions of life, and should plainly bear the stamp of far higher workmanship?

It may be said that natural selection is daily and hourly scrutinising, throughout the world, every variation, even the slightest; rejecting that which is bad, preserving and adding up all that is good; silently and insensibly working, whenever and wherever opportunity offers, at the improvement of each organic being in relation to its organic and inorganic conditions of life. We see nothing of these slow changes in progress, until the hand of time has marked the long lapse of ages, and then so imperfect is our view into long past geological ages, that we only see that the forms of life are now different from what they formerly were.

Sexual Selection. — Inasmuch as peculiarities often appear under domestication in one sex and become hereditarily attached to that sex, the same fact probably occurs under nature, and if so, natural selection will be able to modify one sex in its functional relations to the other sex, or in relation to wholly different habits of life in the two sexes, as is sometimes the case with insects. And this leads me to say a few words on what I call Sexual Selection. This depends, not on a struggle for existence, but on a struggle between the males for possession of the females; the result is not death to the unsuccessful competitor, but few or no offspring. Sexual selection is, therefore, less rigorous than natural selection. Generally, the most vigorous males, those which are best fitted for their places in nature, will leave most progeny. But in many cases, victory will depend not on general vigour, but on having special weapons, confined to the male sex. A hornless stag or spurless cock would have a poor chance of leaving offspring. Sexual selection by always allowing the victor to breed might surely give indomitable courage, length to the spur, and strength to the wing to strike in the spurred leg, as well as the brutal cock-fighter, who knows well that he can improve his breed by careful selection of the best cocks. How low in the scale of nature this law of battle descends, I know not; male alligators have been described as fighting, bellowing, and whirling round, like Indians in a war-dance, for the possession of the females; male salmons have been seen fighting all day long; male stag-beetles often bear wounds from the huge mandibles of other males. The war is, perhaps, severest between the males of polygamous animals, and these seem oftenest provided with special weapons. The males of carnivorous animals are already well armed; though to them and to others, special means of defence may be given through means of sexual selection, as the mane to the lion, the shoulder-pad to the boar, and the hooked jaw to the male salmon; for the shield may be as important for victory, as the sword or spear.

Thus it is, as I believe, that when the males and females of any animal have the same general habits of life, but differ in structure, colour, or ornament, such differences have been mainly caused by sexual selection; that is, individual males have had, in successive generations, some slight advantage over other males, in their weapons, means of defence, or charms; and have transmitted these advantages to their male offspring. Yet, I would not wish to atrribute all such sexual differences to this agency: for we see peculiarities arising and becoming attached to the male sex in our domestic animals (as the wattle in male carriers, horn-like protuberances in the cocks of certain fowls, &c.), which we cannot believe to be either useful to the males in battle, or attractive to the females. We see analogous cases under nature, for instance, the tuft of hair on the breast of the turkey-cock, which can hardly be either useful or ornamental to this bird; — indeed, had the tuft appeared under domestication, it would have been called a monstrosity.

Illustrations of the action of Natural Selection. — In order to make it clear how, as I believe, natural selection acts, I must beg permission to give one or two imaginary illustrations. Let us take the case of a wolf, which preys on various animals, securing some by craft, some by strength, and some by fleetness; and let us suppose that the fleetest prey, a deer for instance, had from any change in the country increased in numbers, or that other prey had decreased in numbers, during that season of the year when the wolf is hardest pressed for food. I can under such circumstances see no reason to doubt that the swiftest and slimmest wolves would have the best chance for surviving, and so be preserved or selected, — provided always that they retained strength to master their prey at this or at some other period of the year, when they might be compelled to prey on other animals. I can see no more reason to doubt this, than that man can improve the fleetness of his greyhounds by careful and methodical selection, or by that unconscious selection which results from each man trying to keep the best dogs without any thought of modifying the breed.

Even without any change in the proportional numbers of the animals on which our wolf preyed, a cub might be born with an innate tendency to pursue certain kinds of prey. Nor can this be thought very improbable; for we often observe great differences in the natural tendencies of our domestic animals; one cat, for instance, taking to catch rats, another mice; one cat, according to Mr. St. John, bringing home winged game, another hares or rabbits, and another hunting on marshy ground and almost nightly catching woodcocks or snipes. The tendency to catch rats rather than mice is known to be inherited. Now, if any slight innate change of habit or of structure benefited an individual wolf, it would have the best chance of surviving and of leaving offspring. Some of its young would probably inherit the same habits or structure, and by the repetition of this process, a new variety might be formed which would either supplant or coexist with the parent-form of wolf. Or, again, the wolves inhabiting a mountainous district, and those frequenting the lowlands, would naturally be forced to hunt different prey; and from the continued preservation of the individuals best fitted for the two sites, two varieties might slowly be formed. These varieties would cross and blend where they met; but to this subject of intercrossing we shall soon have to return. I may add, that, according to Mr. Pierce, there are two varieties of the wolf inhabiting the Catskill Mountains in the United States, one with a light greyhound-like form, which pursues deer, and the other more bulky, with shorter legs, which more frequently attacks the shepherd's flocks.

Let us now take a more complex case. Certain plants excrete a sweet juice, apparently for the sake of eliminating something injurious from their sap: this is effected by glands at the base of the stipules in some Leguminosae, and at the back of the leaf of the common laurel. This juice, though small in quantity, is greedily sought by insects. Let us now suppose a little sweet juice or nectar to be excreted by the inner bases of the petals of a flower. In this case insects in seeking the nectar would get dusted with pollen, and would certainly often transport the pollen from one flower to the stigma of another flower. The flowers of two distinct individuals of the same species would thus get crossed; and the act of crossing, we have good reason to believe (as will hereafter be more fully alluded to), would produce very vigorous seedlings, which consequently would have the best chance of flourishing and surviving. Some of these seedlings would probably inherit the nectar-excreting power. Those individual flowers which had the largest glands or nectaries, and which exreted most nectar, would be oftenest visited by insects, and would be oftenest crossed; and so in the long-run would gain the upper hand. Those flowers, also, which had their stamens and pistils placed, in relation to the size and habits of the particular insects which visited them, so as to favour in any degree the transportal of their pollen from flower to flower, would likewise be favoured or selected. We might have taken the case of insects visiting flowers for the sake of collecting pollen instead of nectar; and as pollen is formed for the sole object of fertilisation, its destruction appears a simple loss to the plant; yet if a little pollen were carried, at first occasionally and then habitually, by the pollen-devouring insects from flower to flower, and a cross thus effected, although nine-tenths of the pollen were destroyed, it might still be a great gain to the plant; and those individuals which produced more and more pollen, and had larger and larger anthers, would be selected.

When our plant, by this process of the continued preservation or natural selection of more and more attractive flowers, had been rendered highly attractive to insects, they would unintentionally on their part, regularly carry pollen from flower to flower; and that they can most effectually do this, I could easily show by many striking instances. I will give only one — not as a very striking case, but as likewise illustrating one step in the separation of the sexes of plants, presently to be alluded to. Some holly-trees bear only male flowers, which have four stamens producing rather a small quantity of pollen, and a rudimentary pistil; other holly-trees bear only female flowers; these have a full-sized pistil and four stamens with shrivelled anthers, in which not a grain of pollen can be detected. Having found a female tree exactly sixty yards from a male tree, I put the stigmas of twenty flowers, taken from different branches, under the microscope, and on all, without exception, there were pollen-grains, and on some a profusion of pollen. As the wind had set for several days from the female to the male tree, the pollen could not thus have been carried. The weather had been cold and boisterous, and therefore not favourable to bees, nevertheless every female flower which I examined had been effectually fertilised by the bees, accidentally dusted with pollen, having flown from tree to tree in search of nectar. But to return to our imaginary case: as soon as the plant had been rendered so highly attractive to insects that pollen was regularly carried from flower to flower, another process might commence. No naturalist doubts the advantage of what has been called the "physiological division of labour;" hence we may believe that it would be advantageous to a plant to produce stamens alone in one flower or on one whole plant, and pistils alone in another flower or on another plant. In plants under culture and placed under new conditions of life, sometimes the male organs and sometimes the female organs become more or less impotent; now if we suppose this to occur in ever so slight a degree under nature, then as pollen is already carried regularly from flower to flower, and as a more complete separation of the sexes of our plant would be advantageous on the principle of the division of labour, individuals with this tendency more and more increased would be continually favoured or selected, until at last a complete separation of the sexes would be effected.

Let us now turn to the nectar-feeding insects in our imaginary case: we may suppose the plant of which we have been slowly increasing the nectar by continued selection, to be a common plant; and that certain insects depended in main part on its nectar for food. I could give many facts, showing how anxious bees are to save time; for instance, their habit of cutting holes and sucking the nectar at the bases of certain flowers, which they can, with a very little more trouble, enter by the mouth. Bearing such facts in mind, I can see no reason to doubt that an accidental deviation in the size and form of the body, or in the curvature and length of the proboscis, &c., far too slight to be appreciated by us, might profit a bee or other insect, so that an individual so characterised would be able to obtain its food more quickly, and so have a better chance of living and leaving descendants. Its descendants would probably inherit a tendency to a similar slight deviation of structure. The tubes of the corollas of the common red and incarnate clovers (Trifolium pratense and incarnatum) do not on a hasty glance appear to differ in length; yet the hive-bee can easily suck the nectar out of the incarnate clover, but not out of the common red clover, which is visited by humble-bees alone; so that whole fields of the red clover offer in vain an abundant supply of precious nectar to the hive-bee. Thus it might be a great advantage to the hive-bee to have a slightly longer or differently constructed proboscis. On the other hand, I have found by experiment that the fertility of clover greatly depends on bees visiting and moving parts of the corolla, so as to push the pollen on to the stigmatic surface. Hence, again, if humble-bees were to become rare in any country, it might be a great advantage to the red clover to have a shorter or more deeply divided tube to its corolla, so that the hive-bee could visit its flowers. Thus I can understand how a flower and a bee might slowly become, either simultaneously or one after the other, modifed and adapted in the most perfect manner to each other, by the continued preservation of individuals presenting mutual and slightly favourable deviations of structure.

I am well aware that this doctrine of natural selection, exemplified in the above imaginary instances, is open to the same objections which were at first urged against Sir Charles Lyell's noble views on "the modern changes of the earth, as illustrative of geology;" but we now very seldom hear the action, for instance, of the coast-waves, called a trifling and insignificant cause, when applied to the excavation of gigantic valleys or to the formation of the longest lines of inland cliffs. Natural selection can act only by the preservation and accumulation of infinitesimally small inherited modifications, each profitable to the preserved being; and as modern geology has almost banished such views as the excavation of a great valley by a single diluvial wave, so will natural selection, if it be a true principle, banish the belief of the continued creation of new organic beings, or of any great and sudden modification in their structure.

※

Circumstances favourable to Natural Selection. — This is an extremely intricate subject. A large amount of inheritable and diversified variability is favourable, but I believe mere individual differences suffice for the work. A large number of individuals, by giving a better chance for the appearance within any given period of profitable variations, will compensate for a lesser amount of variability in each individual, and is, I believe, an extremely important element of success. Though nature grants vast periods of time for the work of natural selection, she does not grant an indefinite period; for as all organic beings are striving, it may be said, to seize on each place in the economy of nature, if any one species does not become modified and improved in a corresponding degree with its competitors, it will soon be exterminated.

In man's methodical selection, a breeder selects for some definite object, and free intercrossing will wholly stop his work. But when many men, without intending to alter the breed, have a nearly common standard of perfection, and all try to get and breed from the best animals, much improvement and modification surely but slowly follow from this unconscious process of selection, notwithstanding a large amount of crossing with inferior animals. Thus it will be in nature; for within a confined area, with some place in its polity not so perfectly occupied as might be, natural selection will always tend to preserve all the individuals varying in the right direction, though in different degrees, so as better to fill up the unoccupied place. But if the area be large, its several districts will almost certainly present different conditions of life; and then if natural selection be modifying and improving a species in the several districts, there will be intercrossing with the other individuals of the same species on the confines of each. And in this case the effects of intercrossing can hardly be counterbalanced by natural selection always tending to modify all the individuals in each district in exactly the same manner to the conditions of each; for in a continuous area, the conditions will generally graduate away insensibly from one district to another.

Isolation, also, is an important element in the process of natural selection. In a confined or isolated area, if not very large, the organic and inorganic conditions of life will generally be in a great degree uniform; so that natural selection will tend to modify all the individuals of a varying species throughout the area in the same manner in relation to the same conditions. Intercrosses, also, with the individuals of the same species, which otherwise would have inhabited the surrounding and differently circumstanced districts, will be prevented. But isolation probably acts more efficiently in checking the immigration of better adapted organisms, after any physical change, such as of climate or elevation of the land, &c.; and thus new places in the natural economy of the country are left open for the old inhabitants to struggle for, and become adapted to, through modifications in their structure and constitution. Lastly, isolation, by checking immigration and consequently competition, will give time for any new variety to be slowly improved; and this may sometimes be of importance in the production of new species. If, however, an isolated area be very small, either from being surrounded by barriers, or from having very peculiar physical conditions, the total number of the individuals supported on it will necessarily be very small; and fewness of individuals will greatly retard the production of new species through natural selection, by decreasing the chance of the appearance of favourable variations.

If we turn to nature to test the truth of these remarks, and look at any small isolated area, such as an oceanic island, although the total number of the species inhabiting it will be found to be small, as we shall see in our chapter on geographical distribution; yet of these species a very large proportion are endemic, — that is, having been produced there, and nowhere else. Hence an oceanic island at first sight seems to have been highly favourable for the production of new species. But we may thus greatly deceive ourselves, for to ascertain whether a small isolated area, or a large open area like a continent, has been most favourable for the production of new organic forms, we ought to make the comparison within equal times; and this we are incapable of doing.

Although I do not doubt that isolation is of considerable importance in the production of new species, on the whole I am inclined to believe that largeness of area is of more importance, more especially in the production of species, which will prove capable of enduring for a long period, and of spreading widely. Throughout a great and open area, not only will there be a better chance of favourable variations arising from the large number of individuals of the same species there supported, but the conditions of life are infinitely complex from the large number of already existing species; and if some of these many species become modified and improved, others will have to be improved in a corresponding degree or they will be exterminated. Each new form, also, as soon as it has been much improved, will be able to spread over the open and continuous area, and will thus come into competition with many others. Hence more new places will be formed, and the competition to fill them will be more severe, on a large than on a small and isolated area. Moreover, great areas, though now continuous, owing to oscillations of level, will often have recently existed in a broken condition, so that the good effects of isolation will generally, to a certain extent, have concurred. Finally, I conclude that, although small isolated areas probably have been in some respects highly favourable for the production of new species, yet that the course of modification will generally have been more rapid on large areas; and what is more important, that the new forms produced on large areas, which already have been victorious over many competitors, will be those that will spread most widely, will give rise to most new varieties and species, and will thus play an important part in the changing history of the organic world.

To sum up the circumstances favourable and unfavourable to natural selection, as far as the extreme intricacy of the subject permits. I conclude, looking to the future, that for terrestrial productions a large continental area, which will probably undergo many oscillations of level, and which consequently will exist for long periods in a broken condition, will be the most favourable for the production of many new forms of life, likely to endure long and to spread widely. For the area will first have existed as a continent, and the inhabitants, at this period numerous in individuals and kinds, will have been subjected to very severe competition. When converted by subsidence into large separate islands, there will still exist many individuals of the same species on each island: intercrossing on the confines of the range of each species will thus be checked: after physical chanes of any kind, immigration will be prevented, so that new places in the polity of each island will have to be filled up by modifications of the old inhabitants; and time will be allowed for the varieties in each to become well modified and perfected. When, by renewed elevation, the islands shall be reconverted into a continental area, there will again be severe competition: the most favoured or improved varieties will be enabled to spread: there will be much extinction of the less improved forms, and the relative proportional numbers of the various inhabitants of the renewed continent will again be changed; and again there will be a fair field for natural selection to improve still further the inhabitants, and thus produce new species.

That natural selection will always act with extreme slowness, I fully admit. Its action depends on there being places in the polity of nature, which can be better occupied by some of the inhabitants of the country undergoing modification of some kind. The existence of such places will often depend on physical changes, which are generally very slow, and on the immigration of better adapted forms having been checked. But the action of natural selection will probably still oftener depend on some of the inhabitants becoming slowly modified; the mutual relations of many of the other inhabitants being thus disturbed. Nothing can be effected, unless favourable variations occur, and variation itself is apparently always a very slow process. The process will often be greatly retarded by free intercrossing. Many will exclaim that these several causes are amply sufficient wholly to stop the action of natural selection. I do not believe so. On the other hand, I do believe that natural selection will always act very slowly, often only at long intervals of time, and generally on only a very few of the inhabitants of the same region at the same time. I further believe, that this very slow, intermittent action of natural selection accords perfectly well with what geology tells us of the rate and manner at which the inhabitants of this world have changed.

Slow though the process of selection may be, if feeble man can do much by his powers of artificial selection, I can see no limit to the amount of change, to the beauty and infinite complexity of the coadaptations between all organic beings, one with another and with their physical conditions of life, which may be effected in the long course of time by nature's power of selection.

Extinction. — This subject will be more fully discussed in our chapter on Geology; but it must be here alluded to from being intimately connected with natural selection. Natural selection acts solely through the preservation of variations in some way advantageous, which consequently endure. But as from the high geometrical powers of increase of all organic beings, each area is already fully stocked with inhabitants, it follows that as each selected and favoured form increases in number, so will the less favoured forms decrease and become rare. Rarity, as geology tells us, is the precursor to extinction. We can, also, see that any form represented by few individuals will, during fluctuations in the seasons or in the number of its enemies, run a good chance of utter extinction. But we may go further than this; for as new forms are continually and slowly being produced, unless we believe that the number of specific forms goes on perpetually and almost indefinitely increasing, numbers inevitably must become extinct. That the number of specific forms has not indefinitely increased, geology shows us plainly; and indeed we can see reason why they should not have thus increased, for the number of places in the polity of nature is not indefinitely great, — not that we have any means of knowing that any one region has as yet got its maximum of species. Probably no region is as yet fully stocked, for at the Cape of Good Hope, where more species of plants are crowded together than in any other quarter of the world, some foreign plants have become naturalised, without causing, as far as we know, the extinction of any natives.

Furthermore, the species which are most numerous in individuals will have the best chance of producing within any given period favourable variations. We have evidence of this, in the facts given in the second chapter, showing that it is the common species which afford the greatest number of recorded varieties, or incipient species. Hence, rare species will be less quickly modified or improved within any given period, and they will consequently be beaten in the race for life by the modified descendants of the commoner species.

From these several considerations I think it inevitably follows, that as new species in the course of time are formed through natural selection, others will become rarer and rarer, and finally extinct. The forms which stand in closest competition with those undergoing modification and improvement, will naturally suffer most. And we have seen in the chapter on the Struggle for Existence that it is the most closely-allied forms, — varieties of the same species, and species of the same genus or of related genera, — which, from having nearly the same structure, constitution, and habits, generally come into the severest competition with each other. Consequently, each new variety or species, during the progress of its formation, will generally press hardest on its nearest kindred, and tend to exterminate them. We see the same process of extermination amongst our domesticated productions, through the selection of improved forms by man. Many curious instances could be given showing how quickly new breeds of cattle, sheep, and other animals, and varieties of flowers, take the place of older and inferior kinds. In Yorkshire, it is historically known that the ancient black cattle were displaced by the long-horns, and that these "were swept away by the short-horns" (I quote the words of an agricultural writer) "as if by some murderous pestilence."

Divergence of Character. — The principle, which I have designated by this term, is of high importance on my theory, and explains, as I believe, several important facts. In the first place, varieties, even strongly-marked ones, though having somewhat of the character of species — as is shown by the hopeless doubts in many cases how to rank them — yet certainly differ from each other far less than do good and distinct species. Nevertheless, according to my view, varieties are species in the process of formation, or are, as I have called them, incipient species. How, then, does the lesser difference between varieties become augmented into the greater difference between species? That this does habitually happen, we must infer from most of the innumerable species throughout nature presenting well-marked differences; whereas varieties, the supposed prototypes and parents of future well-marked species, present slight and ill-defined differences. Mere chance, as we may call it, might cause one variety to differ in some character from its parents, and the offspring of this variety again to differ from its parent in the very same character and in a greater degree; but this alone would never account for so habitual and large an amount of difference as that between varieties of the same species and species of the same genus.

As has always been my practice, let us seek light on this head from our domestic productions. We shall here find something analogous. A fancier is struck by a pigeon having a slightly shorter beak; another fancier is struck by a pigeon having a rather longer beak; and on the acknowledged principle that "fanciers do not and will not admire a medium standard, but like extremes," they both go on (as has actually occurred with tumbler-pigeons) choosing and breeding from birds with longer and longer beaks, or with shorter and shorter beaks. Again, we may suppose that at an early period one man preferred swifter horses; another stronger and more bulky horses. The early differences would be very slight; in the course of time, from the continued selection of swifter horses by some breeders, and of stronger ones by others, the differences would become greater, and would be noted as forming two sub-breeds; finally, after the lapse of centuries, the sub-breeds would become converted into two well-established and distinct breeds. As the differences slowly become greater, the inferior animals with intermediate characters, being neither very swift nor very strong, will have been neglected, and will have tended to disappear. Here, then, we see in man's productions the action of what may be called the principle of divergence, causing differences, at first barely appreciable, steadily to increase, and the breeds to diverge in character both from each other and from their common parent.

But how, it may be asked, can any analogous principle apply in nature? I believe it can and does apply most efficiently, from the simple circumstance that the more diversified the descendants from any one species become in structure, constitution, and habits, by so much will they be better enabled to seize on many and widely diversified places in the polity of nature, and so be enabled to increase in numbers.

We can clearly see this in the case of animals with simple habits. Take the case of a carnivorous quadruped, of which the number that can be supported in any country has long ago arrived at its full average. If its natural powers of increase be allowed to act, it can succeed in increasing (the country not undergoing any change in its conditions) only by its varying descendants seizing on places at present occupied by other animals: some of them, for instance, being enabled to feed on new kinds of prey, either dead or alive; some inhabiting new stations, climbing trees, frequenting water, and some perhaps becoming less carnivorous. The more diversified in habits and structure the descendants of our carnivorous animal became, the more places they would be enabled to occupy. What applies to one animal will apply throughout all time to all animals — that is, if they vary — for otherwise natural selection can do nothing.

The advantage of diversification in the inhabitants of the same region is, in fact, the same as that of the physiological division of labour in the organs of the same individual body — a subject so well elucidated by Milne Edwards. No physiologist doubts that a stomach by being adapted to digest vegetable matter alone, or flesh alone, draws most nutriment from these substances. So in the general economy of any land, the more widely and perfectly the animals and plants are diversified for different habits of life, so will a greater number of individuals be capable of there supporting themselves. A set of animals, with their organisation but little diversified, could hardly compete with a set more perfectly diversified in structure. It may be doubted, for instance, whether the Australian marsupials, which are divided into groups differing but little from each other, and feebly representing, as Mr. Waterhouse and others have remarked, our carnivorous, ruminant, and rodent mammals, could successfully compete with these well-pronounced orders. In the Australian mammals, we see the process of diversification in an early and incomplete stage of development.

Summary of Chapter. — If during the long course of ages and under varying conditions of life, organic beings vary at all in the several parts of their organisation, and I think this cannot be disputed; if there be, owing to the high geometrical powers of increase of each species, at some age, season, or year, a severe struggle for life, and this certainly cannot be disputed; then, considering the infinite complexity of the relations of all organic beings to each other and to their conditions of existence, causing an infinite diversity in structure, constitution, and habits, to be advantageous to them, I think it would be a most extraordinary fact if no variation ever had occurred useful to each being's own welfare, in the same way as so many variations have occurred useful to man. But if variations useful to any organic being do occur, assuredly individuals thus characterised will have the best chance of being preserved in the struggle for life; and from the strong principle of inheritance they will tend to produce offspring similarly characterised. This principle of preservation, I have called, for the sake of brevity, Natural Selection. Natural selection, on the principle of qualities being inherited at corresponding ages, can modify the egg, seed, or young, as easily as the adult. Amongst many animals, sexual selection will give its aid to ordinary selection, by assuring to the most vigorous and best adapted males the greatest number of offspring. Sexual selection will also give characters useful to the males alone, in their struggles with other males.

Whether natural selection has really thus acted in nature, in modifying and adapting the various forms of life to their several conditions and stations, must be judged of by the general tenour and balance of evidence given in the following chapters. But we already see how it entails extinction; and how largely extinction has acted in the world's history, geology plainly declares. Natural selection, also, leads to divergence of character; for more living beings can be supported on the same area the more they diverge in structure, habits, and constitution, of which we see proof by looking at the inhabitants of any small spot or at naturalised productions. Therefore during the modification of the descendants of any one species, and during the incessant struggle of all species to increase in numbers, the more diversified these descendants become, the better will be their chance of succeeding in the battle of life. Thus the small differences distinguishing varieties of the same species will steadily tend to increase till they come to equal the greater differences between species of the same genus, or even of distinct genera.

We have seen that it is the common, the widely-diffused, and widely-ranging species, belonging to the larger genera, which vary most; and these will tend to transmit to their modified offspring that superiority which now makes them dominant in their own countries. Natural selection, as has just been remarked, leads to divergence of character and to much extinction of the less improved and intermediate forms of life. On these principles, I believe, the nature of the affinities of all organic beings may be explained. It is a truly wonderful fact — the wonder of which we are apt to overlook from familiarity — that all animals and all plants throughout all time and space should be related to each other in group subordinate to group, in the manner which we everywhere behold — namely, varieties of the same species most closely related together, species of the same genus less closely and unequally related together, forming sections and sub-genera, species of distinct genera much less closely related, and genera related in different degrees, forming sub-families, families, orders, sub-classes, and classes. The several subordinate groups in any class cannot be ranked in a single file, but seem rather to be clustered round points, and these round other points, and so on in almost endless cycles. On the view that each species has been independently created, I can see no explanation of this great fact in the classification of all organic beings; but, to the best of my judgment, it is explained through inheritance and the complex action of natural selection, entailing extinction and divergence of character, as we have seen illustrated in the diagram.

The affinities of all the beings of the same class have some-times been represented by a great tree. I believe this simile largely speaks the truth. The green and budding twigs may represent existing species; and those produced during each former year may represent the long succession of extinct species. At each period of growth all the growing twigs have tried to branch out on all sides, and to overtop and kill the surrounding twigs and branches, in the same manner as species and groups of species have tried to overmaster other species in the great battle for life. The limbs divided into great branches, and these into lesser and lesser branches, were themselves once, when the tree was small, budding twigs; and this connexion of the former and present buds by ramifying branches may well represent the classification of all extinct and living species in groups subor-dinate to groups. Of the many twigs which flourished when the tree was a mere bush, only two or three, now grown into great branches, yet survive and bear all the other branches; so with the species which lived during long-past geological pe-riods, very few now have living and modified descendants. From the first growth of the tree, many a limb and branch has de-cayed and dropped off; and these lost branches of various sizes may represent those whole orders, families, and genera which have now no living representatives, and which are known to us only from having been found in a fossil state. As we here and there see a thin straggling branch springing from a fork low down in a tree, and which by some chance has been fa-voured and is still alive on its summit, so we occasionally see an animal like the Ornithorhynchus or Lepidosiren, which in some small degree connects by its affinities two large branches of life, and which has apparently been saved from fatal com-petition by having inhabited a protected station. As buds give rise by growth to fresh buds, and these, if vigorous, branch out and overtop on all sides many a feebler branch, so by genera-tion I believe it has been with the great Tree of Life, which fills with its dead and broken branches the crust of the earth, and covers the surface with its ever branching and beautiful ramifications.

Perhaps most of the difficulties Darwin was to experience in gaining acceptance for his thesis that evolution was directed by natural selection are traceable to the largely erroneous views on inheritance expressed here. He hypothecates in this chapter five different sources of inherited variability to fuel the evolutionary process. The first of these, the indirect effect of environment on the reproductive organs, he credits most as the agency initiating variability. As noted previously, it is here that we find a reasonable agreement with our present concept of genetic mutation, since these variations appear spontaneously and unoriented to the needs of the organism. This does not hold, however, for the next two classes of variation, those that allegedly arise either in direct response to the environment, or in response to habit, according to their sustained use and disuse by the organism. The principle here is essentially Lamarckian, of course. A fourth source of variation Darwin refers to as correlation in growth, whereby whole suites of traits are bound together. The student of modern genetics will undoubtedly recognize that this phenomenon encompasses several possible mechanisms, including pleiotropism — a single gene regulating multiple traits — and linkage of genes on the same chromosome. The fifth source of variation, described as a principle of compensation or balance, stems from an analogy with the law of conservation of energy, so that selection for increased development of one trait is believed to lead to atrophication for another. Only the first and fourth of these mechanisms do we recognize today.

An even more formidable misconception, although perhaps less plainly apparent in this chapter, is Darwin's blending interpretation of heredity. Like others of his day, he believed that from its initial appearance the expression of a trait tends to be diminished by half in each successive generation of outbreeding. Thus the parental determinants of heredity are believed to mix inseparably in the offspring, much like pigments of paint mixed in a pot. By contrast, of course, we know now that the units of inheritance — the genes — remain essentially unaltered through descent (save by occasional mutation), without blending to diminish their expression. Darwin's blending theory was to prove eventually quite troublesome, especially when he was pressed to explain how a favorable trait, arising for the first time in a population, might persist there in the face of its tendency toward successive dilution, like a drop of white paint dissolved into a pot of black. If in each generation its expression — and hence its adaptive advantage — were reduced by half, then its eventual establishment by natural selection would be difficult at best. We shall find that much of the future course of Darwin's inquiry was shaped by this misunderstanding.

CHAPTER V

Laws of Variation

I HAVE hitherto sometimes spoken as if the variations — so common and multiform in organic beings under domestication, and in a lesser degree in those in a state of nature — had been due to chance. This, of course, is a wholly incorrect expression, but it serves to acknowledge plainly our ignorance of the cause of each particular variation. Some authors believe it to be as much the function of the reproductive system to produce individual differences, or very slight deviations of structure, as to make the child like its parents. But the much greater variability, as well as the greater frequency of monstrosities, under domestication or cultivation, than under nature, leads me to believe that deviations of structure are in some way due to the nature of the conditions of life, to which the parents and their more remote ancestors have been exposed during several generations. I have remarked in the first chapter — but a long catalogue of facts which cannot be here given would be necessary to show the truth of the remark — that the reproductive system is eminently susceptible to changes in the conditions of life; and to this system being functionally disturbed in the parents, I chiefly attribute the varying or plastic condition of the offspring. The male and female sexual elements seem to be affected before that union takes place which is to form a new being. In the case of "sporting" plants, the bud, which in its earliest condition does not apparently differ essentially from an ovule, is alone affected. But why, because the reproductive system is disturbed, this or that part should vary more or less, we are profoundly ignorant. Nevertheless, we can here and there dimly catch a faint ray of light, and we may feel sure that there must be some cause for each deviation of structure, however slight.

When a variation is of the slightest use to a being, we cannot tell how much of it to attribute to the accumulative action of natural selection, and how much to the conditions of life. Thus, it is well known to furriers that animals of the same species have thicker and better fur the more severe the climate is under which they have lived; but who can tell how much of this difference may be due to the warmest-clad individuals having been favoured and preserved during many generations, and how much to the direct action of the severe climate? for it would appear that climate has some direct action on the hair of our domestic quadrupeds.

Instances could be given of the same variety being produced under conditions of life as different as can well be conceived;

and, on the other hand, of different varieties being produced from the same species under the same conditions. Such facts show how indirectly the conditions of life must act. Again, innumerable instances are known to every naturalist of species keeping true, or not varying at all, although living under the most opposite climates. Such considerations as these incline me to lay very little weight on the direct action of the conditions of life. Indirectly, as already remarked, they seem to play an important part in affecting the reproductive system, and in thus inducing variability; and natural selection will then accumulate all profitable variations, however slight, until they become plainly developed and appreciable by us.

Effects of Use and Disuse. — From the facts alluded to in the first chapter, I think there can be little doubt that use in our domestic animals strengthens and enlarges certain parts, and disuse diminishes them; and that such modifications are inherited. Under free nature, we can have no standard of comparison, by which to judge of the effects of long-continued use or disuse, for we know not the parent-forms; but many animals have structures which can be explained by the effects of disuse.

The eyes of moles and of some burrowing rodents are rudimentary in size, and in some cases are quite covered up by skin and fur. This state of the eyes is probably due to gradual reduction from disuse, but aided perhaps by natural selection. In South America, a burrowing rodent, the tuco-tuco, or Ctenomys, is even more subterranean in its habits than the mole; and I was assured by a Spaniard, who had often caught them, that they were frequently blind; one which I kept alive was certainly in this condition, the cause, as appeared on dissection, having been inflammation of the nictitating membrane. As frequent inflammation of the eyes must be injurious to any animal, and as eyes are certainly not indispensable to animals with subterranean habits, a reduction in their size with the adhesion of the eyelids and growth of fur over them, might in such case be an advantage; and if so, natural selection would constantly aid the effects of disuse.

It is well known that several animals, belonging to the most different classes, which inhabit the caves of Styria and of Kentucky, are blind. In some of the crabs the foot-stalk for the eye remains, though the eye is gone; the stand for the telescope is there, though the telescope with its glasses has been lost. As it is difficult to imagine that eyes, though useless, could be in any way injurious to animals living in darkness, I attribute their loss wholly to disuse. In one of the blind animals, namely, the cave-rat, the eyes are of immense size; and Professor Silliman thought that it regained, after living some days in the light, some slight power of vision. In the same manner as in Madeira the wings of some of the insects have been enlarged, and the wings of others have been reduced by natural selection aided by use and disuse, so in the case of the cave-rat natural selection seems to have struggled with the loss of light and to have increased the size of the eyes; whereas with all the other inhabitants of the caves, disuse by itself seems to have done its work.

On the whole, I think we may conclude that habit, use, and disuse, have, in some cases, played a considerable part in the modification of the constitution, and of the structure of various organs; but that the effects of use and disuse have often been largely combined with, and sometimes overmastered by, the natural selection of innate differences.

Correlation of Growth. — I mean by this expression that the whole organisation is so tied together during its growth and development, that when slight variations in any one part occur, and are accumulated through natural selection, other parts become modified. This is a very important subject, most imperfectly understood. The most obvious case is, that modifications accumulated solely for the good of the young or larva, will, it may safely by concluded, affect the structure of the adult; in the same manner as any malconformation affecting the early embryo, seriously affects the whole organisation of the adult. The several parts of the body which are homologous, and which, at an early embryonic period, are alike, seem liable to vary in an allied manner: we see this in the right and left sides of the body varying in the same manner; in the front and hind legs, and even in the jaws and limbs, varying together, for the lower jaw is believed to be homologous with the limbs. These tendencies, I do not doubt, may be mastered more or less completely by natural selection: thus a family of stags once existed with an antler only on one side; and if this had been of any great use to the breed it might probably have been rendered permanent by natural selection.

The elder Geoffroy and Goethe propounded, at about the same period, their law of compensation or balancement of growth; or, as Goethe expressed it, "in order to spend on one side, nature is forced to economise on the other side." I think this holds true to a certain extent with our domestic productions: if nourishment flows to one part or organ in excess, it rarely flows, at least in excess, to another part; thus it is difficult to get a cow to give much milk and to fatten readily. The same varieties of the cabbage do not yield abundant and nutritious foliage and a copious supply of oil-bearing seeds. When the seeds in our fruits become atrophied, the fruit itself gains largely in size and quality. In our poultry, a large tuft of feathers on the head is generally accompanied by a diminished comb, and a large beard by diminished wattles. With species in a state of nature it can hardly be maintained that the law is of universal application; but many good observers, more especially botanists, believe in its truth. I will not, however, here give any instances, for I see hardly any way of distinguishing between the effects, on the one hand, of a part being largely developed through natural selection and another and adjoining part being reduced by this same process or by disuse, and, on the other hand, the actual withdrawal of nutriment from one part owing to the excess of growth in another and adjoining part.

Distinct species present analogous variations; and a variety of one species often assumes some of the characters of an allied species, or reverts to some of the characters of an early progenitor. — These propositions will be most readily understood by looking to our domestic races. The most distinct breeds of pigeons, in countries most widely apart, present sub-varieties with reversed feathers on the head and feathers on the feet, — characters not possessed by the aboriginal rock-pigeon; these then are analogous variations in two or more distinct races. The frequent presence of fourteen or even sixteen tail-feathers in the pouter, may be considered as a variation representing the normal structure of another race, the fantail. I presume that no one will doubt that all such analogous variations are due to the several races of the pigeon having inherited from a common parent the same constitution and tendency to variation, when acted on by similar unknown influences. In the vegetable kingdom we have a case of analogous variation, in the enlarged stems, or roots as commonly called, of the Swedish turnip and Ruta baga, plants which several botanists rank as varieties produced by cultivation from a common parent: if this be not so, the case will then be one of analogous variation in two so-called distinct species; and to these a third may be added, namely, the common turnip. According to the ordinary view of each species having been independently created, we should have to attribute this similarity in the enlarged stems of these three plants, not to the *vera causa* of community of descent, and a consequent tendency to vary in a like manner, but to three separate yet closely related acts of creation.

No doubt it is a very surprising fact that characters should reappear after having been lost for many, perhaps for hundreds of generations. But when a breed has been crossed only once by some other breed, the offspring occasionally show a tendency to revert in character to the foreign breed for many generations — some say, for a dozen or even a score of generations. After twelve generations, the proportion of blood, to use a common expression, of any one ancestor, is only 1 in 2048; and yet, as we see, it is generally believed that a tendency to reversion is retained by this very small proportion of foreign blood. In a breed which has not been crossed, but in which *both* parents have lost some character which their progenitor possessed, the tendency, whether strong or weak, to reproduce the lost character might be, as was formerly remarked, for all that we can see to the contrary, transmitted for almost any number of generations. When a character which has been lost in a breed reappears after a great number of generations, the most probable hypothesis is, not that the offspring suddenly takes after an ancestor some hundred generations distant, but that in each successive generation there has been a tendency to reproduce the character in question, which at last, under unknown favourable conditions, gains an ascendancy. For instance, it is probable that in each generation of the barb-pigeon, which produces most rarely a blue and black-barred bird, there has been a tendency in each generation in the plumage to assume this colour. This view is hypothetical, but could be supported by some facts; and I can see no more abstract improbability in a tendency to produce any character being inherited for an endless number of generations, than in quite useless or rudimentary organs being, as we all know them to be, thus inherited. Indeed, we may sometimes observe a mere tendency to produce a rudiment inherited: for instance, in the common snap-dragon (Antirrhinum) a rudiment of a fifth stamen so often appears, that this plant must have an inherited tendency to produce it.

❋

Whatever the cause may be of each slight difference in the offspring from their parents — and a cause for each must exist — it is the steady accumulation, through natural selection, of such differences, when beneficial to the individual, that gives rise to all the more important modifications of structure, by which the innumerable beings on the face of this earth are enabled to struggle with each other, and the best adapted to survive.

Here Darwin raises and dispatches a number of questions that he expected his readers to find in his theory. First, he returns to the problem of why species seem often well-defined, and not continuously graduated into one another. Where before he emphasized the role of competition and extinction in accomplishing this delineation, here his explanation centers upon the fragmentation of habitats into different ecological zones. Accordingly, a transitional variety occupying the narrow intermediate band between larger primary zones soon tends to be eliminated by virtue of its small number; small populations contain fewer variations, and therefore fewer potentially adaptive variations, than the more numerous varieties inhabiting the wider adjacent zones. It is interesting to note how this argument seems so little informed by his seminal observations of species differentiation in the Galapagos. In fact, while Darwin discounts the necessity for geographic isolation in establishing new species, he does posit an analogous role for ecological isolation. For the most part, however, the modern position is that geographic isolation is a prerequisite for speciation, and that the effectiveness of ecological isolation alone to inhibit the swamping (by gene exchange) of emergent species differences is doubtful.

It is also in this context that he makes direct assault on the argument for creation by design (see page 139). The intricate perfection of the vertebrate eye had enjoyed some lengthy standing as the best-known illustration of the argument by design, and here Darwin posits the incremental assembly of the eye by natural selection in very early vertebrates, citing a series of arthropod eyes of graduated complexity to furnish his argument support by analogy. Besides, he allows, the vertebrate eye isn't really so perfect after all.

This rhetorical device of challenging unsuccessfully one's own argument worked rather well for Darwin on one notable occasion. On the recommendation of a referee who responded that "Mr. Darwin had so brilliantly surmounted the formidable obstacles which he was honest enough to put in his own path," his publisher, John Murray, agreed to publish the Origin manuscript.[2]

CHAPTER VI

Difficulties on Theory

LONG before having arrived at this part of my work, a crowd of difficulties will have occurred to the reader. Some of them are so grave that to this day I can never reflect on them without being staggered; but, to the best of my judgment, the greater number are only apparent, and those that are real are not, I think, fatal to my theory.

These difficulties and objections may be classed under the following heads: — Firstly, why, if species have descended from other species by insensibly fine gradations, do we not everywhere see innumerable transitional forms? Why is not all nature in confusion instead of the species being, as we see them, well defined?

Secondly, is it possible that an animal having, for instance, the structure and habits of a bat, could have been formed by the modification of some animal with wholly different habits? Can we believe that natural selection could produce, on the one hand, organs of trifling importance, such as the tail of a giraffe, which serves as a fly-flapper, and, on the other hand, organs of such wonderful structure, as the eye, of which we hardly as yet fully understand the inimitable perfection?

Thirdly, can instincts be acquired and modified through natural selection? What shall we say to so marvellous an instinct as that which leads the bee to make cells, which have practically anticipated the discoveries of profound mathematicians?

Fourthly, how can we account for species, when crossed, being sterile and producing sterile offspring, whereas, when varieties are crossed, their fertility is unimpaired?

The two first heads shall be here discussed — Instinct and Hybridism in separate chapters.

On the absence or rarity of transitional varieties. — As natural selection acts solely by the preservation of profitable modifications, each new form will tend in a fully-stocked country to take the place of, and finally to exterminate, its own less improved parent or other less-favoured forms with which it comes into competition. Thus extinction and natural selection will, as we have seen, go hand in hand. Hence, if we look at each species as descended from some other unknown form, both the parent and all the transitional varieties will generally have been exterminated by the very process of formation and perfection of the new form.

But, as by this theory innumerable transitional forms must have existed, why do we not find them embedded in countless numbers in the crust of the earth? It will be much more convenient to discuss this question in the chapter on the Imperfection of the geological record; and I will here only state that I believe the answer mainly lies in the record being incomparably less perfect than is generally supposed; the imperfection of the record being chiefly due to organic beings not inhabiting profound depths of the sea, and to their remains being embedded and preserved to a future age only in masses of sediment sufficiently thick and extensive to withstand an enormous amount of future degradation; and such fossiliferous masses can be accumulated only where much sediment is deposited on the shallow bed of the sea, whilst it slowly subsides. These contingencies will concur only rarely, and after enormously long intervals. Whilst the bed of the sea is stationary or is rising, or when very little sediment is being deposited, there will be blanks in our geological history. The crust of the earth is a vast museum; but the natural collections have been made only at intervals of time immensely remote.

But it may be urged that when several closely-allied species inhabit the same territory we surely ought to find at the present time many transitional forms. Let us take a simple case: in travelling from north to south over a continent, we generally meet at successive intervals with closely allied or representative species, evidently filling nearly the same place in the natural economy of the land. These representative species often meet and interlock; and as the one becomes rarer and rarer, the other becomes more and more frequent, till the one replaces the other. But if we compare these species where they intermingle, they are generally as absolutely distinct from each other in every detail of structure as are specimens taken from the metropolis inhabited by each. By my theory these allied species have descended from a common parent; and during the process of modification, each has become adapted to the conditions of life of its own region, and has supplanted and exterminated its original parent and all the transitional varieties between its past and present states. Hence we ought not to expect at the present time to meet with numerous transitional varieties in each region, though they must have existed there, and may be embedded there in a fossil condition. But in the intermediate region, having intermediate conditions of life, why do we not now find closely-linking intermediate varieties? This difficulty for a long time quite confounded me. But I think it can be in large part explained.

In the first place we should be extremely cautious in inferring, because an area is now continuous, that it has been continuous during a long period. Geology would lead us to believe that almost every continent has been broken up into islands even during the later tertiary periods; and in such islands distinct species might have been separately formed without the possibility of intermediate varieties existing in the intermediate zones. By changes in the form of the land and of climate, marine areas now continuous must often have existed within recent times in a far less continuous and uniform condition than at present. But I will pass over this way of escaping from the difficulty; for I believe that many perfectly defined species have been formed on strictly continuous areas; though I do not doubt that the formerly broken condition of areas now continuous has played an important part in the formation of new species, more especially with freely-crossing and wandering animals.

In looking at species as they are now distributed over a wide area, we generally find them tolerably numerous over a large territory, then becoming somewhat abruptly rarer and rarer on the confines, and finally disappearing. Hence the neutral territory between two representative species is generally narrow in comparison with the territory proper to each. We see the same fact in ascending mountains, and sometimes it is quite remarkable how abruptly, as Alph. De Candolle has observed, a common alpine species disappears. The same fact has been noticed by Forbes in sounding the depths of the sea with the dredge. To those who look at climate and the physical conditions of life as the all-important elements of distribution, these facts ought to cause surprise, as climate and height or depth graduate away insensibly. But when we bear in mind that almost every species, even in its metropolis, would increase immensely in numbers, were it not for other competing species; that nearly all either prey on or serve as prey for others; in short, that each organic being is either directly or indirectly related in the most important manner to other organic beings, we must see that the range of the inhabitants of any country by no means exclusively depends on insensibly changing physical conditions, but in large part on the presence of other species, on which it depends, or by which it is destroyed, or with which it comes into competition; and as these species are already defined objects (however they may have become so), not blending one into another by insensible gradations, the range of any one species, depending as it does on the range of others, will tend to be sharply defined. Moreover, each species on the confines of its range, where it exists in lessened numbers, will, during fluctuations in the number of its enemies or of its prey, or in the seasons, be extremely liable to utter extermination; and thus its geographical range will come to be still more sharply defined.

If I am right in believing that allied or representative species, when inhabiting a continuous area, are generally so distributed that each has a wide range, with a comparatively narrow neutral territory between them, in which they become rather suddenly rarer and rarer; then, as varieties do not essentially differ from species, the same rule will probably apply to both; and if we in imagination adapt a varying species to a very large area, we shall have to adapt two varieties to two large areas, and a third variety to a narrow intermediate zone. The intermediate variety, consequently, will exist in lesser numbers from inhabiting a narrow and lesser area; and practically, as far as I can make out, this rule holds good with varieties in a state of nature. I have met with striking instances of the rule in the case of varieties intermediate between well-marked varieties in the genus Balanus. And it would appear from information given me by Mr. Watson, Dr. Asa Gray, and Mr. Wollaston, that generally when varieties intermediate between two other forms occur, they are much rarer numerically than the forms which they connect. Now, if we may trust these facts and inferences, and therefore conclude that varieties linking two other varieties together have generally existed in lesser numbers than the forms which they connect, then, I think, we can understand why intermediate varieties should not endure for very long periods; — why as a general rule they should be exterminated and disappear, sooner than the forms which they originally linked together.

For any form existing in lesser numbers would, as already remarked, run a greater chance of being exterminated than one existing in large numbers; and in this particular case the intermediate form would be eminently liable to the inroads of closely allied forms existing on both sides of it. But a far more important consideration, as I believe, is that, during the process of further modification, by which two varieties are supposed on my theory to be converted and perfected into two distinct species, the two which exist in larger numbers from inhabiting larger areas, will have a great advantage over the intermediate variety, which exists in smaller numbers in a narrow and intermediate zone. For forms existing in larger numbers will always have a better chance, within any given period, of presenting further favourable variations for natural selection to seize on, than will the rarer forms which exist in lesser numbers. Hence, the more common forms, in the race for life, will tend to beat and supplant the less common forms, for these will be more slowly modified and improved.

To sum up, I believe that species come to be tolerably well-defined objects, and do not at any one period present an inextricable chaos of varying and intermediate links: firstly, because new varieties are very slowly formed, for variation is a very slow process, and natural selection can do nothing until favourable variations chance to occur, and until a place in the natural polity of the country can be better filled by some modification of some one or more of its inhabitants. And such new places will depend on slow changes of climate, or on the occasional immigration of new inhabitants, and, probably, in a still more important degree, on some of the old inhabitants becoming slowly modified, with the new forms thus produced and the old ones acting and reacting on each other. So that, in any one region and at any one time, we ought only to see a few species presenting slight modifications of structure in some degree permanent; and this assuredly we do see.

Secondly, areas now continuous must often have existed within the recent period in isolated portions, in which many forms, more especially amongst the classes which unite for each birth and wander much, may have separately been rendered sufficiently distinct to rank as representative species. In this case, intermediate varieties between the several representative species and their common parent must formerly have existed in each broken portion of the land, but these links will have been supplanted and exterminated during the process of natural selection, so that they will no longer exist in a living state.

Thirdly, when two or more varieties have been formed in different portions of a strictly continuous area, intermediate varieties will, it is probable, at first have been formed in the intermediate zones, but they will generally have had a short duration. For these intermediate varieties will, from reasons already assigned (namely from what we know of the actual distribution of closely allied or representative species, and likewise of acknowledged varieties), exist in the intermediate zones in lesser numbers than the varieties which they tend to connect. From this cause alone the intermediate varieties will be liable to accidental extermination; and during the process of further modification through natural selection, they will almost certainly be beaten and supplanted by the forms which they connect; for these from existing in greater numbers will, in the aggregate, present more variation, and thus be further improved through natural selection and gain further advantages.

Lastly, looking not to any one time, but to all time, if my theory be true, numberless intermediate varieties, linking most closely all the species of the same group together, must assuredly have existed; but the very process of natural selection constantly tends, as has been so often remarked, to exterminate the parent-forms and the intermediate links. Consequently evidence of their former existence could be found only amongst fossil remains, which are preserved, as we shall in a future chapter attempt to show, in an extremely imperfect and intermittent record.

On the origin and transitions of organic beings with peculiar habits and structure. — It has been asked by the opponents of such views as I hold, how, for instance, a land carnivorous animal could have been converted into one with aquatic habits; for how could the animal in its transitional state have subsisted? It would be easy to show that within the same group carnivorous animals exist having every intermediate grade between truly aquatic and strictly terrestrial habits; and as each exists by a struggle for life, it is clear that each is well adapted in its habits to its place in nature. Look at the Mustela vison of North America, which has webbed feet and which resembles an otter in its fur, short legs, and form of tail; during summer this animal dives for and preys on fish, but during the long winter it leaves the frozen waters, and preys like other pole-cats on mice and land animals. If a different case had been taken, and it had been asked how an insectivorous quadruped could possibly have been converted into a flying bat, the question would have been far more difficult, and I could have given no answer. Yet I think such difficulties have very little weight.

Here, as on other occasions, I lie under a heavy disadvantage, for out of the many striking cases which I have collected, I can give only one or two instances of transitional habits and structures in closely allied species of the same genus; and of diversified habits, either constant or occasional, in the same species. And it seems to me that nothing less than a long list of such cases is sufficient to lessen the difficulty in any particular case like that of the bat.

Look at the family of squirrels; here we have the finest gradation from animals with their tails only slightly flattened, and from others, as Sir J. Richardson has remarked, with the posterior part of their bodies rather wide and with the skin on their flanks rather full, to the so-called flying squirrels; and flying squirrels have their limbs and even the base of the tail united by a broad expanse of skin, which serves as a parachute and allows them to glide through the air to an astonishing distance from tree to tree. We cannot doubt that each structure is of use to each kind of squirrel in its own country, by enabling it to escape birds or beasts of prey, or to collect food more quickly, or, as there is reason to believe, by lessening the danger from occasional falls. But it does not follow from this fact that the structure of each squirrel is the best that it is possible to conceive under all natural conditions. Let the climate and vegetation change, let other competing rodents or new beasts of prey immigrate, or old ones become modified, and all analogy would lead us to believe that some at least of the squirrels would decrease in numbers or become exterminated, unless they also became modified and improved in structure in a corresponding manner. Therefore, I can see no difficulty, more especially under changing conditions of life, in the continued preservation of individuals with fuller and fuller flank-membranes, each modification being useful, each being propagated, until by the accumulated effects of this process of natural selection, a perfect so-called flying squirrel was produced.

Now look at the Galeopithecus or flying lemur, which formerly was falsely ranked amongst bats. It has an extremely wide flank-membrane, stretching from the corners of the jaw to the tail, and including the limbs and the elongated fingers: the flank-membrane is, also, furnished with an extensor muscle. Although no graduated links of structure, fitted for gliding through the air, now connect the Galeopithecus with the other Lemuridæ, yet I can see no difficulty in supposing that such links formerly existed, and that each had been formed by the same steps as in the case of the less perfectly gliding squirrels; and that each grade of structure had been useful to its possessor. Nor can I see any insuperable difficulty in further believing it possible that the membrane-connected fingers and fore-arm of the Galeopithecus might be greatly lengthened by natural selection; and this, as far as the organs of flight are concerned, would convert it into a bat. In bats which have the wing-membrane extended from the top of the shoulder to the tail, including the hind-legs, we perhaps see traces of an apparatus originally constructed for gliding through the air rather than for flight.

If about a dozen genera of birds had become extinct or were unknown, who would have ventured to have surmised that birds might have existed which used their wings solely as flappers, like the logger-headed duck (Micropterus of Eyton); as fins in the water and front legs on the land, like the penguin; as sails, like the ostrich; and functionally for no purpose, like the Apteryx. Yet the structure of each of these birds is good for it, under the conditions of life to which it is exposed, for each has to live by a struggle; but it is not necessarily the best possible under all possible conditions. It must not be inferred from these remarks that any of the grades of wing-structure here alluded to, which perhaps may all have resulted from disuse, indicate the natural steps by which birds have acquired their perfect power of flight; but they serve, at least, to show what diversified means of transition are possible.

Seeing that a few members of such water-breathing classes as the Crustacea and Mollusca are adapted to live on the land, and seeing that we have flying birds and mammals, flying insects of the most diversified types, and formerly had flying reptiles, it is conceivable that flying-fish, which now glide far through the air, slightly rising and turning by the aid of their fluttering fins, might have been modified into perfectly winged animals. If this had been effected, who would have ever imagined that in an early transitional state they had been inhabitants of the open ocean, and had used their incipient organs of flight exclusively, as far as we know, to escape being devoured by other fish?

When we see any structure highly perfected for any particular habit, as the wings of bird for flight, we should bear in mind that animals displaying early transitional grades of the structure will seldom continue to exist to the present day, for they will have been supplanted by the very process of perfection through natural selection. Furthermore, we may conclude that transitional grades between structures fitted for very different habits of life will rarely have been developed at an early period in great numbers and under many subordinate forms. Thus, to return to our imaginary illustration of the flying-fish, it does not seem probable that fishes capable of true flight would have been developed under many subordinate forms, for taking prey of many kinds in many ways, on the land and in the water, until their organs of flight had come to a high stage of perfection, so as to have given them a decided advantage over other animals in the battle for life. Hence the chance of discovering species with transitional grades of structure in a fossil condition will always be less, from their having existed in lesser numbers, than in the case of species with fully developed structures.

I will now give two or three instances of diversified and of changed habits in the individuals of the same species. When either case occurs, it would be easy for natural selection to fit the animal, by some modification of its structure, for its changed habits, or exclusively for one of its several different habits. But it is difficult to tell, and immaterial for us, whether habits generally change first and structure afterwards; or whether slight modifications of structure lead to changed habits; both probably often change almost simultaneously. Of cases of changed habits it will suffice merely to allude to that of the many British insects which now feed on exotic plants, or exclusively on artificial substances. Of diversified habits innumerable instances could be given: I have often watched a tyrant flycatcher (Saurophagus sulphuratus) in South America, hovering over one spot and then proceeding to another, like a kestrel, and at other times standing stationary on the margin of water, and then dashing like a kingfisher at a fish. In our own country the larger titmouse (Parus major) may be seen climbing branches, almost like a creeper; it often, like a shrike, kills small birds by blows on the head; and I have many times seen and heard it hammering the seeds of the yew on a branch, and thus breaking them like a nuthatch. In North America the black bear was seen by Hearne swimming for hours with widely open mouth, thus catching, like a whale, insects in the water. Even in so extreme a case as this, if the supply of insects were constant, and if better adapted competitors did not already exist in the country, I can see no difficulty in a race of bears being rendered, by natural selection, more and more aquatic in their structure and habits, with larger and larger mouths, till a creature was produced as monstrous as a whale.

As we sometimes see individuals of a species following habits widely different from those both of their own species and of the other species of the same genus, we might expect, on my theory, that such individuals would occasionally have given rise to new species, having anomalous habits, and with their structure either slightly or considerably modified from that of their proper type. And such instances do occur in nature. Can a more striking instance of adaptation be given than that of a woodpecker for climbing trees and for seizing insects in the chinks of the bark? Yet in North America there are woodpeckers which feed largely on fruit, and others with elongated wings which chase insects on the wing; and on the plains of La Plata, where not a tree grows, there is a woodpecker, which in every essential part of its organisation, even in its colouring, in the harsh tone of its voice, and undulatory flight, told me plainly of its close blood-relationship to our common species; yet it is a woodpecker which never climbs a tree!

He who believes that each being has been created as we now see it, must occasionally have felt surprise when he has met with an animal having habits and structure not at all in agreement. What can be plainer than that the webbed feet of ducks and geese are formed for swimming? yet there are upland geese with webbed feet which rarely or never go near the water; and no one except Audubon has seen the frigate-bird, which has all its four toes webbed, alight on the surface of the sea. On the other hand, grebes and coots are eminently aquatic, although their toes are only bordered by membrane. What seems plainer than that the long toes of grallatores are formed for walking over swamps and floating plants, yet the water-hen is nearly as aquatic as the coot; and the landrail nearly as terrestrial as the quail or partridge. In such cases, and many others could be given, habits have changed without a corresponding change of structure. The webbed feet of the upland goose may be said to have become rudimentary in function, though not in structure. In the frigate-bird, the deeply-scooped membrane between the toes shows that structure has begun to change.

He who believes in separate and innumerable acts of creation will say, that in these cases it has pleased the Creator to cause a being of one type to take the place of one of another type; but this seems to me only restating the fact in dignified language. He who believes in the struggle for existence and in the principle of natural selection will acknowledge that every organic being is constantly endeavouring to increase in numbers; and that if any one being vary ever so little, either in habits or structure, and thus gain an advantage over some other inhabitant of the country, it will seize on the place of that inhabitant, however different it may be from its own place. Hence it will cause him no surprise that there should be geese and frigate-birds with webbed feet, either living on the dry land or most rarely alighting on the water; that there should be long-toed corncrakes living in meadows instead of in swamps; that there should be woodpeckers where not a tree grows; that there should be diving thrushes, and petrels with the habits of auks.

Organs of extreme perfection and complication. — To suppose that the eye, with all its inimitable contrivances for adjusting the focus to different distances, for admitting different amounts of light, and for the correction of spherical and chromatic aberration, could have been formed by natural selection, seems, I freely confess, absurd in the highest possible degree. Yet reason tells me, that if numerous gradations from a perfect and complex eye to one very imperfect and simple, each grade being useful to its possessor, can be shown to exist; if further, the eye does vary ever so slightly, and the variations be inherited, which is certainly the case; and if any variation or modification in the organ be ever useful to an animal under changing conditions of life, then the difficulty of believing that a perfect and complex eye could be formed by natural selection, though insuperable by our imagination, can hardly be considered real. How a nerve comes to be sensitive to light hardly concerns us more than how life itself first originated; but I may remark that several facts make me suspect that any sensitive nerve may be rendered sensitive to light, and likewise to those coarser vibrations of the air which produce sound.

In looking for the gradations by which an organ in any species has been perfected, we ought to look exclusively to its lineal ancestors; but this is scarcely ever possible, and we are forced in each case to look to species of the same group, that is to the collateral descendants from the same original parent-form, in order to see what gradations are possible, and for the chance of some gradations having been transmitted from the earlier stages of descent, in an unaltered or little altered condition. Amongst existing Vertebrata, we find but a small amount of gradation in the structure of the eye, and from fossil species we can learn nothing on this head. In this great class we should probably have to descend far beneath the lowest known fossiliferous stratum to discover the earlier stages, by which the eye has been perfected.

In the Articulata we can commence a series with an optic nerve merely coated with pigment, and without any other mechanism; and from this low stage, numerous gradations of structure, branching off in two fundamentally different lines, can be shown to exist, until we reach a moderately high stage of perfection. In certain crustaceans, for instance, there is a double cornea, the inner one divided into facets, within each of which there is a lens-shaped swelling. In other crustaceans the transparent cones which are coated by pigment, and which properly act only by excluding lateral pencils of light, are convex at their upper ends and must act by convergence; and at their lower ends there seems to be an imperfect vitreous substance. With these facts, here far too briefly and imperfectly given, which show that there is much graduated diversity in the eyes of living crustaceans, and bearing in mind how small the number of living animals is in proportion to those which have become extinct, I can see no very great difficulty (not more than in the case of many other structures) in believing that natural selection has converted the simple apparatus of an optic nerve merely coated with pigment and invested by transparent membrane, into an optical instrument as perfect as is possessed by any member of the great Articulate class.

He who will go thus far, if he find on finishing this treatise that large bodies of facts, otherwise inexplicable, can be explained by the theory of descent, ought not to hesitate to go further, and to admit that a structure even as perfect as the eye of an eagle might be formed by natural selection, although in this case he does not know any of the transitional grades. His reason ought to conquer his imagination; though I have felt the difficulty far too keenly to be surprised at any degree of hesitation in extending the principle of natural selection to such startling lengths.

It is scarcely possible to avoid comparing the eye to a telescope. We know that this instrument has been perfected by the long-continued efforts of the highest human intellects; and we naturally infer that the eye has been formed by a somewhat analogous process. But may not this inference be presumptuous? Have we any right to assume that the Creator works by intellectual powers like those of man? If we must compare the eye to an optical instrument, we ought in imagination to take a thick layer of transparent tissue, with a nerve sensitive to light beneath, and then suppose every part of this layer to be continually changing slowly in density, so as to separate into layers of different densities and thicknesses, placed at different distances from each other, and with the surfaces of each layer slowly changing in form. Further we must suppose that there is a power always intently watching each slight accidental alteration in the transparent layers; and carefully selecting each alteration which, under varied circumstances, may in any way, or in any degree, tend to produce a distincter image. We must suppose each new state of the instrument to be multiplied by the million; and each to be preserved till a better be produced, and then the old ones to be destroyed. In living bodies, variation will cause the slight alterations, generation will multiply them almost infinitely, and natural selection will pick out with unerring skill each improvement. Let this process go on for millions on millions of years; and during each year on millions of individuals of many kinds; and may we not believe that a living optical instrument might thus be formed as superior to one of glass, as the works of the Creator are to those of man?

If it could be demonstrated that any complex organ existed, which could not possibly have been formed by numerous, successive, slight modifications, my theory would absolutely break down. But I can find out no such case. No doubt many organs exist of which we do not know the transitional grades, more especially if we look to much-isolated species, round which, according to my theory, there has been much extinction. Or again, if we look to an organ common to all the members of a large class, for in this latter case the organ must have been first formed at an extremely remote period, since which all the many members of the class have been developed; and in order to discover the early transitional grades through which the organ has passed, we should have to look to very ancient ancestral forms, long since become extinct.

We should be extremely cautious in concluding that an organ could not have been formed by transitional gradations of some kind. Numerous cases could be given amongst the lower animals of the same organ performing at the same time wholly distinct functions; thus the alimentary canal respires, digests, and excretes in the larva of the dragon-fly and in the fish Cobites. In the Hydra, the animal may be turned inside out, and the exterior surface will then digest and the stomach respire. In such cases natural selection might easily specialise, if any advantage were thus gained, a part or organ, which had performed two functions, for one function alone, and thus wholly change its nature by insensible steps. Two distinct organs sometimes perform simultaneously the same function in the same individual; to give one instance, there are fish with gills or branchiæ that breathe the air dissolved in the water, at the same time that they breathe free air in their swimbladders, this latter organ having a ductus pneumaticus for its supply, and being divided by highly vascular partitions. In these cases, one of the two organs might with ease be modified and perfected so as to perform all the work by itself, being aided during the process of modification by the other organ; and then this other organ might be modified for some other and quite distinct purpose, or be quite obliterated.

The illustration of the swimbladder in fishes is a good one, because it shows us clearly the highly important fact that an organ originally constructed for one purpose, namely flotation, may be converted into one for a wholly different purpose, namely respiration. The swimbladder has, also, been worked in as an accessory to the auditory organs of certain fish, or, for I do not know which view is now generally held, a part of the auditory apparatus has been worked in as a complement to the swimbladder. All physiologists admit that the swimbladder is homologous, or "ideally similar," in position and structure with the lungs of the higher vertebrate animals: hence there seems to me to be no great difficulty in believing that natural selection has actually converted a swimbladder into a lung, or organ used exclusively for respiration.

Although in many cases it is most difficult to conjecture by what transitions an organ could have arrived at its present state; yet, considering that the proportion of living and known forms to the extinct and unknown is very small, I have been astonished how rarely an organ can be named, towards which no transitional grade is known to lead. The truth of this remark is indeed shown by that old canon in natural history of "Natura non facit saltum." We meet with this admission in the writings of almost every experienced naturalist; or, as Milne Edwards has well expressed it, nature is prodigal in variety, but niggard in innovation. Why, on the theory of Creation, should this be so? Why should all the parts and organs of many independent beings, each supposed to have been separately created for its proper place in nature, be so invariably linked together by graduated steps? Why should not Nature have taken a leap from structure to structure? On the theory of natural selection, we can clearly understand why she should not; for natural selection can act only by taking advantage of slight successive variations; she can never take a leap, but must advance by the shortest and slowest steps.

Organs of little apparent importance. — As natural selection acts by life and death, — by the preservation of individuals with any favourable variation, and by the destruction of those with any unfavourable deviation of structure, — I have sometimes felt much difficulty in understanding the origin of simple parts, of which the importance does not seem sufficient to cause the preservation of successively varying individuals. I have sometimes felt as much difficulty, though of a very different kind, on this head, as in the case of an organ as perfect and complex as the eye.

In the first place, we are much too ignorant in regard to the whole economy of any one organic being, to say what slight modifications would be of importance or not. . . . The tail of the giraffe looks like an artificially constructed fly-flapper; and it seems at first incredible that this could have been adapted for its present purpose by successive slight modifications, each better and better, for so trifling an object as driving away flies; yet we should pause before being too positive even in this case, for we know that the distribution and existence of cattle and other animals in South America absolutely depends on their power of resisting the attacks of insects: so that individuals which could by any means defend themselves from these small enemies would be able to range into new pastures and thus gain a great advantage. It is not that the larger quadrupeds are actually destroyed (except in some rare cases) by the flies, but they are incessantly harassed and their strength reduced, so that they are more subject to disease, or not so well enabled in a coming dearth to search for food, or to escape from beasts of prey.

Organs now of trifling importance have probably in some cases been of high importance to an early progenitor, and, after having been slowly perfected at a former period, have been transmitted in nearly the same state, although now become of very slight use; and any actually injurious deviations in their structure will always have been checked by natural selection. Seeing how important an organ of locomotion the tail is in most aquatic animals, its general presence and use for many purposes in so many land animals, which in their lungs or modified swim-bladders betray their aquatic origin, may perhaps be thus accounted for. A well-developed tail having been formed in an aquatic animal, it might subsequently come to be worked in for all sorts of purposes, as a fly-flapper, an organ of prehension, or as an aid in turning, as with the dog, though the aid must be slight, for the hare, with hardly any tail, can double quickly enough.

The foregoing remarks lead me to say a few words on the protest lately made by some naturalists, against the utilitarian doctrine that every detail of structure has been produced for the good of its possessor. They believe that very many structures have been created for beauty in the eyes of man, or for mere variety. This doctrine, if true, would be absolutely fatal to my theory. Yet I fully admit that many structures are of no direct use to their possessors. Physical conditions probably have had some little effect on structure, quite independently of any good thus gained. Correlation of growth has no doubt played a most important part, and a useful modification of one part will often have entailed on other parts diversified changes of no direct use. So again characters which formerly were useful, or which formerly had arisen from correlation of growth, or from other unknown cause, may reappear from the law of reversion, though now of no direct use. The effects of sexual selection, when displayed in beauty to charm the females, can be called useful only in rather a forced sense. But by far the most important consideration is that the chief part of the organisation of every being is simply due to inheritance; and consequently, though each being assuredly is well fitted for its place in nature, many structures now have no direct relation to the habits of life of each species. Thus, we can hardly believe that the webbed feet of the upland goose or of the frigate-bird are of special use to these birds; we cannot believe that the same bones in the arm of the monkey, in the fore leg of the horse, in the wing of the bat, and in the flipper of the seal, are of special use to these animals. We may safely attribute these structures to inheritance. But to the progenitor of the upland goose and of the frigate-bird, webbed feet no doubt were as useful as they now are to the most aquatic of existing birds. So we may believe that the progenitor of the seal had not a flipper, but a foot with five toes fitted for walking or grasping; and we may further venture to believe that the several bones in the limbs of the monkey, horse, and bat, which have been inherited from a common progenitor, were formerly of more special use to that progenitor, or its progenitors, than they now are to these animals having such widely diversified habits. Therefore we may infer that these several bones might have been acquired through natural selection, subjected formerly, as now, to the several laws of inheritance, reversion, correlation of growth, &c. Hence every detail of structure in every living creature (making some little allowance for the direct action of physical conditions) may be viewed, either as having been of special use to some ancestral form, or as being now of special use to the descendants of this form — either directly, or indirectly through the complex laws of growth.

Natural selection cannot possibly produce any modification in any one species exclusively for the good of another species; though throughout nature one species incessantly takes advantage of, and profits by, the structure of another. But natural selection can and does often produce structures for the direct injury of other species, as we see in the fang of the adder, and in the ovipositor of the ichneumon, by which its eggs are deposited in the living bodies of other insects. If it could be proved that any part of the structure of any one species had been formed for the exclusive good of another species, it would annihilate my theory, for such could not have been produced through natural selection. Although many statements may be found in works on natural history to this effect, I cannot find even one which seems to me of any weight. It is admitted that the rattlesnake has a poison-fang for its own defence and for the destruction of its prey; but some authors suppose that at the same time this snake is furnished with a rattle for its own injury, namely, to warn its prey to escape. I would almost as soon believe that the cat curls the end of its tail when preparing to spring, in order to warn the doomed mouse. But I have not space here to enter on this and other such cases.

Natural selection will never produce in a being anything injurious to itself, for natural selection acts solely by and for the good of each. No organ will be formed, as Paley has remarked, for the purpose of causing pain or for doing an injury to its possessor. If a fair balance be struck between the good and evil caused by each part, each will be found on the whole advantageous. After the lapse of time, under changing conditions of life, if any part comes to be injurious, it will be modified; or if it be not so, the being will become extinct, as myriads have become extinct.

Natural selection tends only to make each organic being as perfect as, or slightly more perfect than, the other inhabitants of the same country with which it has to struggle for existence. And we see that this is the degree of perfection attained under nature. The endemic productions of New Zealand, for instance, are perfect one compared with another; but they are now rapidly yielding before the advancing legions of plants and animals introduced from Europe. Natural selection will not produce absolute perfection, nor do we always meet, as far as we can judge, with this high standard under nature. The correction for the aberration of light is said, on high authority, not to be perfect even in that most perfect organ, the eye. If our reason leads us to admire with enthusiasm a multitude of inimitable contrivances in nature, this same reason tells us, though we may easily err on both sides, that some other contrivances are less perfect. Can we consider the sting of the wasp or of the bee as perfect, which, when used against many attacking animals, cannot be withdrawn, owing to the backward serratures, and so inevitably causes the death of the insect by tearing out its viscera?

If we look at the sting of the bee, as having originally existed in a remote progenitor as a boring and serrated instrument, like that in so many members of the same great order, and which has been modified but not perfected for its present purpose, with the poison originally adapted to cause galls subsequently intensified, we can perhaps understand how it is that the use of the sting should so often cause the insect's own death: for if on the whole the power of stinging be useful to the community, it will fulfil all the requirements of natural selection, though it may cause the death of some few members. If we admire the truly wonderful power of scent by which the males of many insects find their females, can we admire the production for this single purpose of thousands of drones, which are utterly useless to the community for any other end, and which are ultimately slaughtered by their industrious and sterile sisters? It may be difficult, but we ought to admire the savage instinctive hatred of the queen-bee, which urges her instantly to destroy the young queens her daughters as soon as born, or to perish herself in the combat; for undoubtedly this is for the good of the community; and maternal love or maternal hatred, though the latter fortunately is most rare, is all the same to the inexorable principle of natural selection. If we admire the several ingenious contrivances, by which the flowers of the orchis and of many other plants are fertilised through insect agency, can we consider as equally perfect the elaboration by our fir-trees of dense clouds of pollen, in order that a few granules may be wafted by a chance breeze on to the ovules?

Summary of Chapter. — We have in this chapter discussed some of the difficulties and objections which may be urged against my theory. Many of them are very grave; but I think that in the discussion light has been thrown on several facts, which on the theory of independent acts of creation are utterly obscure. We have seen that species at any one period are not indefinitely variable, and are not linked together by a multitude of intermediate gradations, partly because the process of natural selection will always be very slow, and will act, at any one time, only on a very few forms; and partly because the very process of natural selection almost implies the continual supplanting and extinction of preceding and intermediate gradations. Closely allied species, now living on a continuous area, must often have been formed when the area was not continuous, and when the conditions of life did not insensibly graduate away from one part to another. When two varieties are formed in two districts of a continuous area, an intermediate variety will often be formed, fitted for an intermediate zone; but from reasons assigned, the intermediate variety will usually exist in lesser numbers than the forms which it connects; consequently the two latter, during the course of further modification, from existing in greater numbers, will have a great advantage over the less numerous intermediate variety, and will thus generally succeed in supplanting and exterminating it.

We have seen in this chapter how cautious we should be in concluding that the most different habits of life could not graduate into each other; that a bat, for instance, could not have been formed by natural selection from an animal which at first could only glide through the air.

We have seen that a species may under new conditions of life change its habits, or have diversified habits, with some habits very unlike those of its nearest congeners. Hence we can understand, bearing in mind that each organic being is trying to live wherever it can live, how it has arisen that there are upland geese with webbed feet, ground woodpeckers, diving thrushes, and petrels with the habits of auks.

Although the belief that an organ so perfect as the eye could have been formed by natural selection, is more than enough to stagger any one; yet in the case of any organ, if we know of a long series of gradations in complexity, each good for its possessor, then, under changing conditions of life, there is no logical impossibility in the acquirement of any conceivable degree of perfection through natural selection. In the cases in which we know of no intermediate or transitional states, we should be very cautious in concluding that none could have existed, for the homologies of many organs and their intermediate states show that wonderful metamorphoses in function are at least possible. For instance, a swimbladder has apparently been converted into an air-breathing lung. The same organ having performed simultaneously very different functions, and then having been specialised for one function; and two very distinct organs having performed at the same time the same function, the one having been perfected whilst aided by the other, must often have largely facilitated transitions.

We are far too ignorant, in almost every case, to be enabled to assert that any part or organ is so unimportant for the welfare of a species, that modifications in its structure could not have been slowly accumulated by means of natural selection. But we may confidently believe that many modifications, wholly due to the laws of growth, and at first in no way advantageous to a species, have been subsequently taken advantage of by the still further modified descendants of this species. We may, also, believe that a part formerly of high importance has often been retained (as the tail of an aquatic animal by its terrestrial descendants), though it has become of such small importance that it could not, in its present state, have been acquired by natural selection, — a power which acts solely by the preservation of profitable variations in the struggle for life.

Natural selection will produce nothing in one species for the exclusive good or injury of another; though it may well produce parts, organs, and excretions highly useful or even indispensable, or highly injurious to another species, but in all cases at the same time useful to the owner. Natural selection in each well-stocked country, must act chiefly through the competition of the inhabitants one with another, and consequently will produce perfection, or strength in the battle for life, only according to the standard of that country. Hence the inhabitants of one country, generally the smaller one, will often yield, as we see they do yield, to the inhabitants of another and generally larger country. For in the larger country there will have existed more individuals, and more diversified forms, and the competition will have been severer, and thus the standard of perfection will have been rendered higher. Natural selection will not necessarily produce absolute perfection; nor, as far as we can judge by our limited faculties, can absolute perfection be everywhere found.

On the theory of natural selection we can clearly understand the full meaning of that old canon in natural history, "Natura non facit saltum." This canon, if we look only to the present inhabitants of the world, is not strictly correct, but if we include all those of past times, it must by my theory be strictly true.

It is generally acknowledged that all organic beings have been formed on two great laws — Unity of Type, and the Conditions of Existence. By unity of type is meant that fundamental agreement in structure, which we see in organic beings of the same class, and which is quite independent of their habits of life. On my theory, unity of type is explained by unity of descent. The expression of conditions of existence, so often insisted on by the illustrious Cuvier, is fully embraced by the principle of natural selection. For natural selection acts by either now adapting the varying parts of each being to its organic and inorganic conditions of life; or by having adapted them during long-past periods of time: the adaptations being aided in some cases by use and disuse, being slightly affected by the direct action of the external conditions of life, and being in all cases subjected to the several laws of growth. Hence, in fact, the law of the Conditions of Existence is the higher law; as it includes, through the inheritance of former adaptations, that of Unity of Type.

In this chapter Darwin directs his theory to the evolution of inherited behavior, arguing that the same principles apply here as to the evolution of corporeal structures. Most fascinating is his evolutionary analysis of sterile worker castes among ants. Their very existence would seem on its face to contradict all precepts of natural selection, since obviously such organisms die without issue. But here we witness Darwin's great improvisational genius, for he cleverly shifts the level at which selection operates from the individual to the family, and finds it advantageous both to parents and their fertile offspring to produce workers adapted to meet the needs of their community. In fact, this explanation based upon kinship anticipates astonishingly well the detailed solution to the problem given a century later under the rubric of kin-selection theory, a theory explaining the evolution of those behaviors without direct reproductive benefit to the individual organism.

CHAPTER VII

Instinct

THE subject of instinct might have been worked into the previous chapters; but I have thought that it would be more convenient to treat the subject separately, especially as so wonderful an instinct as that of the hive-bee making its cells will probably have occurred to many readers, as a difficulty sufficient to overthrow my whole theory. I must premise, that I have nothing to do with the origin of the primary mental powers, any more than I have with that of life itself. We are concerned only with the diversities of instinct and of the other mental qualities of animals within the same class.

I will not attempt any definition of instinct. It would be easy to show that several distinct mental actions are commonly embraced by this term; but every one understands what is meant, when it is said that instinct impels the cuckoo to migrate and to lay her eggs in other birds' nests. An action, which we ourselves should require experience to enable us to perform, when performed by an animal, more especially by a very young one, without any experience, and when performed by many individuals in the same way, without their knowing for what purpose it is performed, is usually said to be instinctive. But I could show that none of these characters of instinct are universal. A little dose, as Pierre Huber expresses it, of judgment or reason, often comes into play, even in animals very low in the scale of nature.

If we suppose any habitual action to become inherited — and I think it can be shown that this does sometimes happen — then the resemblance between what originally was a habit and an instinct becomes so close as not to be distinguished. If Mozart, instead of playing the pianoforte at three years old with wonderfully little practice, had played a tune with no practice at all, he might truly be said to have done so instinctively. But it would be the most serious error to suppose that the greater number of instincts have been acquired by habit in one generation, and then transmitted by inheritance to succeeding generations. It can be clearly shown that the most wonderful instincts with which we are acquainted, namely, those of the hive-bee and of many ants, could not possibly have been thus acquired.

It will be universally admitted that instincts are as important as corporeal structure for the welfare of each species, under its present conditions of life. Under changed conditions of life, it is at least possible that slight modifications of instinct might be profitable to a species; and if it can be shown that instincts do vary ever so little, then I can see no difficulty in natural selection preserving and continually accumulating variations of instinct to any extent that may be profitable. It is thus, as I believe, that all the most complex and wonderful instincts have originated. As modifications of corporeal structure arise from, and are increased by, use or habit, and are diminished or lost by disuse, so I do not doubt it has been with instincts. But I believe that the effects of habit are of quite subordinate importance to the effects of the natural selection of what may be called accidental variations of instincts; — that is of variations produced by the same unknown causes which produce slight deviations of bodily structure.

No complex instinct can possibly be produced through natural selection, except by the slow and gradual accumulation of numerous, slight, yet profitable, variations. Hence, as in the case of corporeal structures, we ought to find in nature, not the actual transitional gradations by which each complex instinct has been acquired — for these could be found only in the lineal ancestors of each species — but we ought to find in the collateral lines of descent some evidence of such gradations; or we ought at least to be able to show that gradations of some kind are possible; and this we certainly can do. I have been surprised to find, making allowance for the instincts of animals having been but little observed except in Europe and North America, and for no instinct being known amongst extinct species, how very generally gradations, leading to the most complex instincts, can be discovered. The canon of "Natura non facit saltum" applies with almost equal force to instincts as to bodily organs. Changes of instinct may sometimes be facilitated by the same species having different instincts at different periods of life, or at different seasons of the year, or when placed under different circumstances, &c.; in which case either one or the other instinct might be preserved by natural selection. And such instances of diversity of instinct in the same species can be shown to occur in nature.

No doubt many instincts of very difficult explanation could be opposed to the theory of natural selection, — cases, in which we cannot see how an instinct could possibly have originated; cases, in which no intermediate gradations are known to exist; cases of instinct of apparently such trifling importance, that they could hardly have been acted on by natural selection; cases of instincts almost identically the same in animals so remote in the scale of nature, that we cannot account for their similarity by inheritance from a common parent, and must therefore believe that they have been acquired by independent acts of natural selection. I will not here enter on these several cases, but will confine myself to one special difficulty, which at first appeared to me insuperable, and actually fatal to my whole theory. I allude to the neuters or sterile females in insect-communities: for these neuters often differ widely in instinct and in structure from both the males and fertile females, and yet, from being sterile, they cannot propagate their kind.

The subject well deserves to be discussed at great length, but I will here take only a single case, that of working or sterile ants. How the workers have been rendered sterile is a difficulty; but not much greater than that of any other striking modification of structure; for it can be shown that some insects and other articulate animals in a state of nature occasionally become sterile; and if such insects had been social, and it had been profitable to the community that a number should have been annually born capable of work, but incapable of procreation, I can see no very great difficulty in this being effected by natural selection. But I must pass over this preliminary difficulty. The great difficulty lies in the working ants differing widely from both the males and the fertile females in structure, as in the shape of the thorax and in being destitute of wings and sometimes of eyes, and in instinct. As far as instinct alone is concerned, the prodigious difference in this respect between the workers and the perfect females would have been far better exemplified by the hive-bee. If a working ant or other neuter insect had been an animal in the ordinary state, I should have unhesitatingly assumed that all its characters had been slowly acquired through natural selection; namely, by an individual having been born with some slight profitable modification of structure, this being inherited by its offspring, which again varied and were again selected, and so onwards. But with the working ant we have an insect differing greatly from its parents, yet absolutely sterile; so that it could never have transmitted successively acquired modifications of structure or instinct to its progeny. It may well be asked how is it possible to reconcile this case with the theory of natural selection?

First, let it be remembered that we have innumerable instances, both in our domestic productions and in those in a state of nature, of all sorts of differences of structure which have become correlated to certain ages, and to either sex. We have differences correlated not only to one sex, but to that short period alone when the reproductive system is active, as in the nuptial plumage of many birds, and in the hooked jaws of the male salmon. We have even slight differences in the horns of different breeds of cattle in relation to an artificially imperfect state of the male sex; for oxen of certain breeds have longer horns than in other breeds, in comparison with the horns of the bulls or cows of these same breeds. Hence I can see no real difficulty in any character having become correlated with the sterile condition of certain members of insect-communities: the difficulty lies in understanding how such correlated modifications of structure could have been slowly accumulated by natural selection.

This difficulty, though appearing insuperable, is lessened, or, as I believe, disappears, when it is remembered that selection may be applied to the family, as well as to the individual, and may thus gain the desired end. Thus, a well-flavoured vegetable is cooked, and the individual is destroyed; but the horticulturist sows seeds of the same stock, and confidently expects to get nearly the same variety; breeders of cattle wish the flesh and fat to be well marbled together; the animal has been slaughtered, but the breeder goes with confidence to the same family. I have such faith in the powers of selection, that I do not doubt that a breed of cattle, always yielding oxen with extraordinarily long horns, could be slowly formed by carefully watching which individual bulls and cows, when matched, produced oxen with the longest horns; and yet no one ox could ever have propagated its kind. Thus I believe it has been with social insects: a slight modification of structure, or instinct, correlated with the sterile condition of certain members of the community, has been advantageous to the community: consequently the fertile males and females of the same community flourished, and transmitted to their fertile offspring a tendency to produce sterile members having the same modification. And I believe that this process has been repeated, until that prodigious amount of difference between the fertile and sterile females of the same species has been produced, which we see in many social insects.

But we have not as yet touched on the climax of the difficulty; namely, the fact that the neuters of several ants differ, not only from the fertile females and males, but from each other, sometimes to an almost incredible degree, and are thus divided into two or even three castes. The castes, moreover, do not generally graduate into each other, but are perfectly well defined; being as distinct from each other, as are any two species of the same genus, or rather as any two genera of the same family. Thus in Eciton, there are working and soldier neuters, with jaws and instincts extraordinarily different: in Cryptocerus, the workers of one caste alone carry a wonderful sort of shield on their heads, the use of which is quite unknown: in the Mexican Myrmecocystus, the workers of one caste never leave the nest; they are fed by the workers of another caste, and they have an enormously developed abdomen which secretes a sort of honey, supplying the place of that excreted by the aphides, or the domestic cattle as they may be called, which our European ants guard or imprison.

It will indeed be thought that I have an overweening confidence in the principle of natural selection, when I do not admit that such wonderful and well-established facts at once annihilate my theory. In the simpler case of neuter insects all of one caste or of the same kind, which have been rendered by natural selection, as I believe to be quite possible, different from the fertile males and females, — in this case, we may safely conclude from the analogy of ordinary variations, that each successive, slight, profitable modification did not probably at first appear in all the individual neuters in the same nest, but in a few alone; and that by the long-continued selection of the fertile parents which produced most neuters with the profitable modification, all the neuters ultimately came to have the desired character. On this view we ought occasionally to find neuter-insects of the same species, in the same nest, presenting gradations of structure; and this we do find, even often, considering how few neuter-insects out of Europe have been carefully examined. Mr. F. Smith has shown how surprisingly the neuters of several British ants differ from each other in size and sometimes in colour; and that the extreme forms can sometimes be perfectly linked together by individuals taken out of the same nest: I have myself compared perfect gradations of this kind. It often happens that the larger or the smaller sized workers are the most numerous; or that both large and small are numerous, with those of an intermediate size scanty in numbers.

Thus, as I believe, the wonderful fact of two distinctly defined castes of sterile workers existing in the same nest, both widely different from each other and from their parents, has originated. We can see how useful their production may have been to a social community of insects, on the same principle that the division of labour is useful to civilised man. As ants work by inherited instincts and by inherited tools or weapons, and not by acquired knowledge and manufactured instruments, a perfect division of labour could be effected with them only by the workers being sterile; for had they been fertile, they would have intercrossed, and their instincts and structure would have become blended. And nature has, as I believe, effected this admirable division of labour in the communities of ants, by the means of natural selection. But I am bound to confess, that, with all my faith in this principle, I should never have anticipated that natural selection could have been efficient in so high a degree, had not the case of these neuter insects convinced me of the fact. I have, therefore, discussed this case, at some little but wholly insufficient length, in order to show the power of natural selection, and likewise because this is by far the most serious special difficulty, which my theory has encountered. The case, also, is very interesting, as it proves that with animals, as with plants, any amount of modification in structure can be effected by the accumulation of numerous, slight, and as we must call them accidental, variations, which are in any manner profitable, without exercise or habit having come into play. For no amount of exercise, or habit, or volition, in the utterly sterile members of a community could possibly have affected the structure or instincts of the fertile members, which alone leave descendants. I am surprised that no one has advanced this demonstrative case of neuter insects, against the well-known doctrine of Lamarck.

This is the last of the chapters devoted to the mechanisms by which new species arise. Again Darwin discounts the orthodox taxonomic criterion for identifying the species — its reproductive isolation from all others — and claims that the demarcation between varieties and species is in fact quite indistinct. Thus, he argues, between a state of complete interfertility among varieties of a species and a state of complete intersterility among species themselves we find a continuous gradation in degree. Moreover, that the degree of reproductive isolation between species seems to increase with their greater taxonomic dissimilarity strengthens his evolutionary interpretation while it weighs against the view of species as products of special creation.

It is also clear from this chapter that Darwin saw the development of reproductive isolation between diverging species only as a coincidental accompaniment to their differentiation on other grounds, since he found no basis for its direct establishment by natural selection. Wallace did not share this view, and recent opinion remains somewhat divided on the matter, since there is no unanimity whether the formation of new species actually does depend on the founding of isolating mechanisms that enhance the individual organism's fitness. Still, as Wallace claimed, it is advantageous for individuals to mate within their own spe-

cies and not to squander their reproductive energies, since poorly adapted or infertile hybrid offspring are usually the best results to be expected from such ecumenical gestures. Thus, natural selection does foster directly the establishment of those isolating mechanisms, such as different courtship behaviors, which early in the breeding process minimize the likelihood of such fruitless encounters.

CHAPTER VIII

Hybridism

THE view generally entertained by naturalists is that species, when intercrossed, have been specially endowed with the quality of sterility, in order to prevent the confusion of all organic forms. This view certainly seems at first probable, for species within the same country could hardly have kept distinct had they been capable of crossing freely. The importance of the fact that hybrids are very generally sterile, has, I think, been much underrated by some late writers. On the theory of natural selection the case is especially important, inasmuch as the sterility of hybrids could not possibly be of any advantage to them, and therefore could not have been acquired by the continued preservation of successive profitable degrees of sterility. I hope, however, to be able to show that sterility is not a specially acquired or endowed quality, but is incidental on other acquired differences.

In treating this subject, two classes of facts, to a large extent fundamentally different, have generally been confounded together; namely, the sterility of two species when first crossed, and the sterility of the hybrids produced from them.

Pure species have of course their organs of reproduction in

a perfect condition, yet when intercrossed they produce either few or no offspring. Hybrids, on the other hand, have their reproductive organs functionally impotent, as may be clearly seen in the state of the male element in both plants and animals; though the organs themselves are perfect in structure, as far as the microscope reveals. In the first case the two sexual elements which go to form the embryo are perfect; in the second case they are either not at all developed, or are imperfectly developed. This distinction is important, when the cause of the sterility, which is common to the two cases, has to be considered. The distinction has probably been slurred over, owing to the sterility in both cases being looked on as a special endowment, beyond the province of our reasoning powers.

The fertility of varieties, that is of the forms known or believed to have descended from common parents, when intercrossed, and likewise the fertility of their mongrel offspring, is, on my theory, of equal importance with the sterility of species; for it seems to make a broad and clear distinction between varieties and species.

It is certain, on the one hand, that the sterility of various species when crossed is so different in degree and graduates away so insensibly, and, on the other hand, that the fertility of pure species is so easily affected by various circumstances, that for all practical purposes it is most difficult to say where perfect fertility ends and sterility begins. I think no better evidence of this can be required than that the two most experienced observers who have ever lived, namely, Kölreuter and Gärtner, should have arrived at diametrically opposite conclusions in regard to the very same species. It is also most instructive to compare — but I have not space here to enter on details — the evidence advanced by our best botanists on the question whether certain doubtful forms should be ranked as species or varieties, with the evidence from fertility adduced by different hybridisers, or by the same author, from experiments made during different years. It can thus be shown that neither sterility nor fertility affords any clear distinction between species and varieties; but that the evidence from this source graduates away, and is doubtful in the same degree as in the evidence derived from other constitutional and structural differences.

In regard to animals, much fewer experiments have been carefully tried than with plants. If our systematic arrangements can be trusted, that is if the genera of animals are as distinct from each other, as are the genera of plants, then we may infer that animals more widely separated in the scale of nature can be more easily crossed than in the case of plants; but the hybrids themselves are, I think, more sterile. I doubt whether any case of a perfectly fertile hybrid animal can be considered as thoroughly well authenticated.

A doctrine which originated with Pallas has been largely accepted by modern naturalists; namely, that most of our domestic animals have descended from two or more aboriginal species, since commingled by intercrossing. On this view, the aboriginal species must either at first have produced quite fertile hybrids, or the hybrids must have become in subsequent generations quite fertile under domestication. This latter alternative seems to me the most probable, and I am inclined to believe in its truth, although it rests on no direct evidence. I believe, for instance, that our dogs have descended from several wild stocks; yet, with perhaps the exception of certain indigenous domestic dogs of South America, all are quite fertile together; and analogy makes me greatly doubt, whether the several aboriginal species would at first have freely bred together and have produced quite fertile hybrids. So again there is reason to believe that our European and the humped Indian cattle are quite fertile together; but from facts communicated to me by Mr. Blyth, I think they must be considered as distinct species. On this view of the origin of many of our domestic animals, we must either give up the belief of the almost universal sterility of distinct species of animals when crossed; or we must look at sterility, not as an indelible characteristic, but as one capable of being removed by domestication.

Finally, looking to all the ascertained facts on the intercrossing of plants and animals, it may be concluded that some degree of sterility, both in first crosses and in hybrids, is an extremely general result; but that it cannot, under our present state of knowledge, be considered as absolutely universal.

Fertility of Varieties when crossed, and of their Mongrel offspring. — It may be urged, as a most forcible argument, that there must be some essential distinction between species and varieties, and that there must be some error in all the foregoing remarks, inasmuch as varieties, however much they may differ from each other in external appearance, cross with perfect facility, and yield perfectly fertile offspring. I fully admit that this is almost invariably the case. But if we look to varieties produced under nature, we are immediately involved in hopeless difficulties; for if two hitherto reputed varieties be found in any degree sterile together, they are at once ranked by most naturalists as species. For instance, the blue and red pimpernel, the primrose and cowslip, which are considered by many of our best botanists as varieties, are said by Gärtner not to be quite fertile when crossed, and he consequently ranks them as undoubted species. If we thus argue in a circle, the fertility of all varieties produced under nature will assuredly have to be granted.

I have as yet spoken as if the varieties of the same species were invariably fertile when intercrossed. But it seems to me impossible to resist the evidence of the existence of a certain amount of sterility in the few following cases, which I will briefly abstract. The evidence is at least as good as that from which we believe in the sterility of a multitude of species. The evidence is, also, derived from hostile witnesses, who in all other cases consider fertility and sterility as safe criterions of specific distinction. Gärtner kept during several years a dwarf kind of maize with yellow seeds, and a tall variety with red seeds, growing near each other in his garden; and although these plants have separated sexes, they never naturally crossed. He then fertilised thirteen flowers of the one with the pollen of the other; but only a single head produced any seed, and this one head produced only five grains. Manipulation in this case could not have been injurious, as the plants have separated sexes. No one, I believe, has suspected that these varieties of maize are distinct species; and. it is important to notice that the hybrid plants thus raised were themselves *perfectly* fertile; so that even Gärtner did not venture to consider the two varieties as specifically distinct.

Kölreuter, whose accuracy has been confirmed by every subsequent observer, has proved the remarkable fact, that one variety of the common tobacco is more fertile, when crossed with a widely distinct species, than are the other varieties. He experimentised on five forms, which are commonly reputed to be varieties, and which he tested by the severest trial, namely, by reciprocal crosses, and he found their mongrel offspring perfectly fertile. But one of these five varieties, when used either as father or mother, and crossed with the Nicotiana glutinosa, always yielded hybrids not so sterile as those which were produced from the four other varieties when crossed with N. glutinosa. Hence the reproductive system of this one variety must have been in some manner and in some degree modified.

From these facts; from the great difficulty of ascertaining the infertility of varieties in a state of nature, for a supposed variety if infertile in any degree would generally be ranked as species; from man selecting only external characters in the production of the most distinct domestic varieties, and from not wishing or being able to produce recondite and functional differences in the reproductive system; from these several considerations and facts, I do not think that the very general fertility of varieties can be proved to be of universal occurrence, or to form a fundamental distinction between varieties and species. The general fertility of varieties does not seem to me sufficient to overthrow the view which I have taken with respect to the very general, but not invariable, sterility of first crosses and of hybrids, namely, that it is not a special endowment, but is incidental on slowly acquired modifications, more especially in the reproductive systems of the forms which are crossed.

In this and the next chapter Darwin looks to geology for substantiation of his theory. What he finds, at least as he reveals in this chapter, is somewhat equivocal testimony. On one hand, the vast tracts of time his account requires are confirmed in the Lyellian reading he gives here to the earth's record. On the other hand, its fossils do not present the continuous sequence of transitional species to connect extinct forms with those extant; indeed, the fossil record seems contrarily discontinuous, marked at times by the abrupt emergence of related groups that lack obvious antecedents.

By Darwin's account, the recalcitrance of paleontology to offer support rests mainly on the imperfection of the earth's record. Thus he attributes the fossil gaps to incomplete geological sampling, to interruptions in the conditions favoring preserva-

tion, to the subsequent erosion of earlier strata, and so forth. But the sudden appearance and dispersal of related forms — birds, for example — he attributes to the explosive proliferation of species into previously unoccupied niches as they capitalize rapidly on a characteristic — wings, in this case — established slowly and incrementally in their ancestral lineage. We hold much the same explanation today, terming the phenomenon adaptive radiation. *The fossil evidence most resistant to his evolutionary interpretation is the sudden appearance of already complex organisms in the Silurian deposits, without obvious living precursors in the underlying strata. In fact, simple multicellular and even unicellular organisms have been identified in much earlier strata, but only quite recently.*

CHAPTER IX

On the Imperfection of the Geological Record

In the sixth chapter I enumerated the chief objections which might be justly urged against the views maintained in this volume. Most of them have now been discussed. One, namely the distinctness of specific forms, and their not being blended together by innumerable transitional links, is a very obvious difficulty. I assigned reasons why such links do not commonly occur at the present day, under the circumstances apparently most favourable for their presence, namely on an extensive and continuous area with graduated physical conditions. I endeavoured to show, that the life of each species depends in a more important manner on the presence of other already defined organic forms, than on climate; and, therefore, that the really governing conditions of life do not graduate away quite insensibly like heat or moisture. I endeavoured, also, to show that intermediate varieties, from existing in lesser numbers than the forms which they connect, will generally be beaten out and exterminated during the course of further modification and improvement. The main cause, however, of innumerable intermediate links not now occurring everywhere throughout nature depends on the very process of natural selection, through which new varieties continually take the places of and exterminate their parent-forms. But just in proportion as this process of extermination has acted on an enormous scale, so must the number of intermediate varieties, which have formerly existed on the earth, be truly enormous. Why then is not every geological formation and every stratum full of such intermediate links? Geology assuredly does not reveal any such finely graduated organic chain; and this, perhaps, is the most obvious and gravest objection which can be urged against my theory. The explanation lies, as I believe, in the extreme imperfection of the geological record.

In the first place it should always be borne in mind what sort of intermediate forms must, on my theory, have formerly existed. I have found it difficult, when looking at any two species, to avoid picturing to myself, forms *directly* intermediate between them. But this is a wholly false view; we should always look for forms intermediate between each species and a common but unknown progenitor; and the progenitor will generally have differed in some respects from all its modified descendants. To give a simple illustration: the fantail and pouter pigeons have both descended from the rock-pigeon; if we possessed all the intermediate varieties which have ever existed, we should have an extremely close series between both and the rock-pigeon; but we should have no varieties directly intermediate between the fantail and pouter; none, for instance, combining a tail somewhat expanded with a crop somewhat enlarged, the characteristic features of these two breeds. These two breeds, moreover, have become so much modified, that if we had no historical or indirect evidence regarding their origin, it would not have been possible to have determined from a mere comparison of their structure with that of the rock-pigeon, whether they had descended from this species or from some other allied species, such as C. oenas.

So with natural species, if we look to forms very distinct, for instance to the horse and tapir, we have no reason to suppose that links ever existed directly intermediate between them, but between each and an unknown common parent. The common parent will have had in its whole organisation much general resemblance to the tapir and to the horse; but in some points of structure may have differed considerably from both, even perhaps more than they differ from each other. Hence in all such cases, we should be unable to recognise the parent-form of any two or more species, even if we closely compared the structure of the parent with that of its modified descendants, unless at the same time we had a nearly perfect chain of the intermediate links.

It is just possible by my theory, that one of two living forms might have descended from the other; for instance, a horse from a tapir; and in this case *direct* intermediate links will have existed between them. But such a case would imply that one form had remained for a very long period unaltered, whilst its descendants had undergone a vast amount of change; and the principle of competition between organism and organism, between child and parent, will render this a very rare event; for in all cases the new and improved forms of life will tend to supplant the old and unimproved forms.

By the theory of natural selection all living species have been connected with the parent-species of each genus, by differences not greater than we see between the varieties of the same species at the present day; and these parent-species, now generally extinct, have in their turn been similarly connected with more ancient species; and so on backwards, always converging to the common ancestor of each great class. So that the number of intermediate and transitional links, between all living and extinct species, must have been inconceivably great. But assuredly, if this theory be true, such have lived upon this earth.

On the lapse of Time. — Independently of our not finding fossil remains of such infinitely numerous connecting links, it may be objected, that time will not have sufficed for so great an amount of organic change, all changes having been effected very slowly through natural selection. It is hardly possible for me even to recall to the reader, who may not be a practical geologist, the facts leading the mind feebly to comprehend the lapse of time. He who can read Sir Charles Lyell's grand work on the Principles of Geology, which the future historian will recognise as having produced a revolution in natural science, yet does not admit how incomprehensibly vast have been the past periods of time, may at once close this volume. Not that it suffices to study the Principles of Geology, or to read special treatises by different observers on separate formations, and to mark how each author attempts to give an inadequate idea of the duration of each formation or even each stratum. A man must for years examine for himself great piles of superimposed strata, and watch the sea at work grinding down old rocks and making fresh sediment, before he can hope to comprehend anything of the lapse of time, the monuments of which we see around us.

It is good to wander along lines of sea-coast, when formed of moderately hard rocks, and mark the process of degradation. The tides in most cases reach the cliffs only for a short time twice a day, and the waves eat into them only when they are charged with sand or pebbles; for there is reason to believe that pure water can effect little or nothing in wearing away rock. At last the base of the cliff is undermined, huge fragments fall down, and these remaining fixed, have to be worn away, atom by atom, until reduced in size they can be rolled about by the waves, and then are more quickly ground into pebbles, sand, or mud. But how often do we see along the bases of retreating cliffs rounded boulders, all thickly clothed by marine productions, showing how little they are abraded and how seldom they are rolled about! Moreover, if we follow for a few miles any line of rocky cliff, which is undergoing degradation, we find that it is only here and there, along a short length or round a promontory, that the cliffs are at the present time suffering. The appearance of the surface and the vegetation show that elsewhere years have elapsed since the waters washed their base.

He who most closely studies the action of the sea on our shores, will, I believe, be most deeply impressed with the slowness with which rocky coasts are worn away. The observations on this head by Hugh Miller, and by that excellent observer Mr. Smith of Jordan Hill, are most impressive. With the mind thus impressed, let any one examine beds of conglomerate many thousand feet in thickness, which, though probably formed at a quicker rate than many other deposits, yet, from being formed of worn and rounded pebbles, each of which bears the stamp of time, are good to show how slowly the mass has been accumulated. Let him remember Lyell's profound remark, that the thickness and extent of sedimentary formations are the result and measure of the degradation which the earth's crust has elsewhere suffered. And what an amount of degradation is implied by the sedimentary deposits of many countries! Professor Ramsay has given me the maximum thickness, in most cases from actual measurement, in a few cases from estimate, of each formation in different parts of Great Britain; and this is the result: —

	Feet.
Palaeozoic strata (not including igneous beds)	57,154
Secondary strata	13,190
Tertiary strata	2,240

— making altogether 72,584 feet; that is, very nearly thirteen and three-quarters British miles. Some of these formations, which are represented in England by thin beds, are thousands of feet in thickness on the Continent. Moreover, between each successive formation, we have, in the opinion of most geologists, enormously long blank periods. So that the lofty pile of sedimentary rocks in Britain gives but an inadequate idea of the time which has elapsed during their accumulation; yet what time this must have consumed! Good observers have estimated that sediment is deposited by the great Mississippi river at the rate of only 600 feet in a hundred thousand years. This estimate may be quite erroneous; yet, considering over what wide spaces very fine sediment is transported by the currents of the sea, the process of accumulation in any one area must be extremely slow.

I am tempted to give one other case, the well-known one of the denudation of the Weald. Though it must be admitted that the denudation of the Weald has been a mere trifle, in comparison with that which has removed masses of our palaeozoic strata, in parts ten thousand feet in thickness, as shown in Prof. Ramsay's masterly memoir on this subject. Yet it is an admirable lesson to stand on the North Downs and to look at the distant South Downs; for, remembering that at no great distance to the west the northern and southern escarpments meet and close, one can safely picture to oneself the great dome of rocks which must have covered up the Weald within so limited a period as since the latter part of the Chalk formation. The distance from the northern to the southern Downs is about 22 miles, and the thickness of the several formations is on an average about 1100 feet, as I am informed by Prof. Ramsay. But if, as some geologists suppose, a range of older rocks underlies the Weald, on the flanks of which the overlying sedimentary deposits might have accumulated in thinner masses than elsewhere, the above estimate would be erroneous; but this source of doubt probably would not greatly affect the estimate as applied to the western extremity of the district. If, then, we knew the rate at which the sea commonly wears away a line of cliff of any given height, we could measure the time requisite to have denuded the Weald. This, of course, cannot be done; but we may, in order to form some crude notion on the subject, assume that the sea would eat into cliffs 500 feet in height at the rate of one inch in a century. This will at first appear much too small an allowance; but it is the same as if we were to assume a cliff one yard in height to be eaten back along a whole line of coast at the rate of one yard in nearly every twenty-two years. I doubt whether any rock, even as soft as chalk, would yield at this rate excepting on the most exposed coasts; though no doubt the degradation of a lofty cliff would be more rapid from the breakage of the fallen fragments. On the other hand, I do not believe that any line of coast, ten or twenty miles in length, ever suffers degradation at the same time along its whole indented length; and we must remember that almost all strata contain harder layers or nodules, which from long resisting attrition form a breakwater at the base. Hence, under ordinary circumstances, I conclude that for a cliff 500 feet in height, a denudation of one inch per century for the whole length would be an ample allowance. At this rate, on the above data, the denudation of the Weald must have required 306,662,400 years; or say three hundred million years.

The action of fresh water on the gently inclined Wealden district, when upraised, could hardly have been great, but it would somewhat reduce the above estimate. On the other hand, during oscillations of level, which we know this area has undergone, the surface may have existed for millions of years as land, and thus have escaped the action of the sea: when deeply submerged for perhaps equally long periods, it would, likewise, have escaped the action of the coast-waves. So that in all probability a far longer period than 300 million years has elapsed since the latter part of the Secondary period.

I have made these few remarks because it is highly important for us to gain some notion, however imperfect, of the lapse of years. During each of these years, over the whole world, the land and the water has been peopled by hosts of living forms. What an infinite number of generations, which the mind cannot grasp, must have succeeded each other in the long roll of years! Now turn to our richest geological museums, and what a paltry display we behold!

On the poorness of our Palaeontological collections. — That our palaeontological collections are very imperfect, is admitted by every one. The remark of that admirable palaeontologist, the late Edward Forbes, should not be forgotten, namely, that numbers of our fossil species are known and named from single and often broken specimens, or from a few specimens collected on some one spot. Only a small portion of the surface of the earth has been geologically explored, and no part with sufficient care, as the important discoveries made every year in Europe prove. No organism wholly soft can be preserved. Shells and bones will decay and disappear when left on the bottom of the sea, where sediment is not accumulating. I believe we are continually taking a most erroneous view, when we tacitly admit to ourselves that sediment is being deposited over nearly the whole bed of the sea, at a rate sufficiently quick to embed and preserve fossil remains. Throughout an enormously large proportion of the ocean, the bright blue tint of the water bespeaks its purity. The many cases on record of a formation conformably covered, after an enormous interval of time, by another and later formation, without the underlying bed having suffered in the interval any wear and tear, seem explicable only on the view of the bottom of the sea not rarely lying for ages in an unaltered condition. The remains which do become embedded, if in sand or gravel, will when the beds are upraised, generally be dissolved by the percolation of rain-water. I suspect that but few of the very many animals which live on the beach between high and low watermark are preserved.

But the imperfection in the geological record mainly results from another and more important cause than any of the foregoing; namely, from the several formations being separated from each other by wide intervals of time. When we see the formations tabulated in written works, or when we follow them in nature, it is difficult to avoid believing that they are closely consecutive. But we know, for instance, from Sir R. Murchison's great work on Russia, what wide gaps there are in that country between the superimposed formations; so it is in North America, and in many other parts of the world. The most skilful geologist, if his attention had been exclusively confined to these large territories, would never have suspected that during the periods which were blank and barren in his own country, great piles of sediment, charged with new and peculiar forms of life, had elsewhere been accumulated. And if in each separate territory, hardly any idea can be formed of the length of time which has elapsed between the consecutive formations, we may infer that this could nowhere be ascertained. The frequent and great changes in the mineralogical composition of consecutive formations, generally implying great changes in the geography of the surrounding lands, whence the sediment has been derived, accords with the belief of vast intervals of time having elapsed between each formation.

All geological facts tell us plainly that each area has undergone numerous slow oscillations of level, and apparently these oscillations have affected wide spaces. Consequently formations rich in fossils and sufficiently thick and extensive to resist subsequent degradation may have been formed over wide spaces during periods of subsidence, but only where the supply of sediment was sufficient to keep the sea shallow and to embed and preserve the remains before they had time to decay. On the other hand, as long as the bed of the sea remained stationary, *thick* deposits could not have been accumulated in the shallow parts, which are the most favourable to life. Still less could this have happened during the alternate periods of elevation; or, to speak more accurately, the beds which were then accumulated will have been destroyed by being upraised and brought within the limits of the coast-action.

Thus the geological record will almost necessarily be rendered intermittent. I feel much confidence in the truth of these views, for they are in strict accordance with the general principles inculcated by Sir C. Lyell; and E. Forbes independently arrived at a similar conclusion.

One remark is here worth a passing notice. During periods of elevation the area of the land and of the adjoining shoal parts of the sea will be increased, and new stations will often be formed; — all circumstances most favourable, as previously explained, for the formation of new varieties and species; but during such periods there will generally be a blank in the geological record. On the other hand, during subsidence, the inhabited area and number of inhabitants will decrease (excepting the productions on the shores of a continent when first broken up into an archipelago), and consequently during subsidence, though there will be much extinction, fewer new varieties or species will be formed; and it is during these very periods of subsidence, that our great deposits rich in fossils have been accumulated. Nature may almost be said to have guarded against the frequent discovery of her transitional or linking forms.

If then, there be some degree of truth in these remarks, we have no right to expect to find in our geological formations, an infinite number of those fine transitional forms, which on my theory assuredly have connected all the past and present species of the same group into one long and branching chain of life. We ought only to look for a few links, some more closely, some more distantly related to each other; and these links, let them be ever so close, if found in different stages of the same formation, would, by most palaeontologists, be ranked as distinct species. But I do not pretend that I should ever have suspected how poor a record of the mutations of life, the best preserved geological section presented, had not the difficulty of our not discovering innumerable transitional links between the species which appeared at the commencement and close of each formation, pressed so hardly on my theory.

On the sudden appearance of whole groups of Allied Species.— The abrupt manner in which whole groups of species suddenly appear in certain formations has been urged by several palaeontologists, for instance, by Agassiz, Pictet, and by none more forcibly than by Professor Sedgwick, as a fatal objection to the belief in the transmutation of species. If numerous species, belonging to the same genera or families, have really started into life all at once, the fact would be fatal to the theory of descent with slow modification through natural selection. For the development of a group of forms, all of which have descended from some one progenitor, must have been an extremely slow process; and the progenitors must have lived long ages before their modified descendants. But we continually overrate the perfection of the geological record, and falsely infer, because certain genera or families have not been found beneath a certain stage, that they did not exist before that stage. We continually forget how large the world is, compared with the area over which our geological formations have been carefully examined; we forget that groups of species may elsewhere have long existed and have slowly multiplied before they invaded the ancient archipelagoes of Europe and of the United States. We do not make due allowance for the enormous intervals of time, which have probably elapsed between our consecutive formations, — longer perhaps in some cases than the time required for the accumulation of each formation. These intervals will have given time for the multiplication of species from some one or some few parent-forms; and in the succeeding formation such species will appear as if suddenly created.

I may here recall a remark formerly made, namely that it might require a long succession of ages to adapt an organism to some new and peculiar line of life, for instance to fly through the air; but that when this had been effected, and a few species had thus acquired a great advantage over other organisms, a comparatively short time would be necessary to produce many divergent forms, which would be able to spread rapidly and widely throughout the world.

From these and similar considerations, but chiefly from our ignorance of the geology of other countries beyond the confines of Europe and the United States; and from the revolution in our palaeontological ideas on many points, which the discoveries of even the last dozen years have effected, it seems to me to be about as rash in us to dogmatize on the succession of organic beings throughout the world, as it would be for a naturalist to land for five minutes on some one barren point in Australia, and then to discuss the number and range of its productions.

On the sudden appearance of groups of Allied Species in the lowest known fossiliferous strata. — There is another and allied difficulty, which is much graver. I allude to the manner in which numbers of species of the same group suddenly appear in the lowest known fossiliferous rocks. Most of the arguments which have convinced me that all the existing species of the same group have descended from one progenitor apply with nearly equal force to the earliest known species. For instance, I cannot doubt that all the Silurian trilobites have descended from some one crustacean, which must have lived long before the Silurian age, and which probably differed greatly from any known animal. Some of the most ancient Silurian animals, as the Nautilus, Lingula, &c., do not differ much from living species; and it cannot on my theory be supposed, that these old species were the progenitors of all the species of the orders to which they belong, for they do not present characters in any degree intermediate between them. If, moreover, they had been the progenitors of these orders, they would almost certainly have been long ago supplanted and exterminated by their numerous and improved descendants.

Consequently, if my theory be true, it is indisputable that before the lowest Silurian stratum was deposited, long periods elapsed, as long as, or probably far longer than, the whole interval from the Silurian age to the present day; and that during these vast, yet quite unknown, periods of time, the world swarmed with living creatures.

To the question why we do not find records of these vast primordial periods, I can give no satisfactory answer. Several of the most eminent geologists, with Sir R. Murchison at their head, are convinced that we see in the organic remains of the lowest Silurian stratum the dawn of life on this planet. Other highly competent judges, as Lyell and the late E. Forbes, dispute this conclusion. We should not forget that only a small portion of the world is known with accuracy.

The several difficulties here discussed, namely our not finding in the successive formations infinitely numerous transitional links between the many species which now exist or have existed; the sudden manner in which whole groups of species appear in our European formations; the almost entire absence, as at present known, of fossiliferous formations beneath the Silurian strata, are all undoubtedly of the gravest nature. We

177

see this in the plainest manner by the fact that all the most eminent palaeontologists, namely Cuvier, Owen, Agassiz, Barrande, Falconer, E. Forbes, &c., and all our greatest geologists, as Lyell, Murchison, Sedgwick, &c., have unanimously, often vehemently, maintained the immutability of species. But I have reason to believe that one great authority, Sir Charles Lyell, from further reflexion entertains grave doubts on this subject. I feel how rash it is to differ from these great authorities, to whom, with others, we owe all our knowledge. Those who think the natural geological record in any degree perfect, and who do not attach much weight to the facts and arguments of other kinds given in this volume, will undoubtedly at once reject my theory. For my part, following out Lyell's metaphor, I look at the natural geological record, as a history of the world imperfectly kept, and written in a changing dialect; of this history we possess the last volume alone, relating only to two or three countries. Of this volume, only here and there a short chapter has been preserved; and of each page, only here and there a few lines. Each word of the slowly-changing language, in which the history is supposed to be written, being more or less different in the interrupted succession of chapters, may represent the apparently abruptly changed forms of life, entombed in our consecutive, but widely separated, formations. On this view, the difficulties above discussed are greatly diminished, or even disappear.

Here the facts of paleontology fall readily into place for Darwin. Fossil species undergo different rates of change, about what would be expected were natural selection governing evolution independently in each lineage. Fossil lineages regularly appear, proliferate, and flourish for a time, and finally relinquish the stage to better-adapted successors, entirely in accord with a vi-

sion of nature ordered by competitive principles. Fossil forms tend to fill the intervals between living species and genera and even families, consistent with a view of evolution as a successive branching process. And inasmuch as the living faunas of the continents are distinctive, so do their fossil faunas mirror this same distinctiveness, an observation Darwin sets forth as "the succession of types." Finally, the fossil record suggests a tendency for later forms to be more complex than their antecedents, lending a generally progressive aspect to evolution. Note how tentatively Darwin broaches this point, no doubt to distance his own account from those incipient stirrings toward progress with which Lamarck and Chambers had earlier encumbered their evolutionary theories.

CHAPTER X

On the Geological Succession of Organic Beings

LET US now see whether the several facts and rules relating to the geological succession of organic beings better accord with the common view of the immutability of species, or with that of their slow and gradual modification, through descent and natural selection.

New species have appeared very slowly, one after another, both on the land and in the waters. Lyell has shown that it is hardly possible to resist the evidence on this head in the case of the several tertiary stages; and every year tends to fill up the blanks between them, and to make the percentage system of lost and new forms more gradual. In some of the most recent beds, though undoubtedly of high antiquity if measured by years, only one or two species are lost forms, and only one or two are new forms, having here appeared for the first time, either locally, or, as far as we know, on the face of the earth. If we may trust the observations of Philippi in Sicily, the successive changes in the marine inhabitants of that island have been many and most gradual. The secondary formations are more broken; but, as Bronn has remarked, neither the appearance nor disappearance of their many now extinct species has been simultaneous in each separate formation.

Species of different genera and classes have not changed at the same rate, or in the same degree. In the oldest tertiary beds a few living shells may still be found in the midst of a multitude of extinct forms. Falconer has given a striking instance of a similar fact, in an existing crocodile associated with many strange and lost mammals and reptiles in the sub-Himalayan deposits. The Silurian Lingula differs but little from the living species of this genus; whereas most of the other Silurian Molluscs and all the Crustaceans have changed greatly. The productions of the land seem to change at a quicker rate than those of the sea, of which a striking instance has lately been observed in Switzerland. There is some reason to believe that organisms, considered high in the scale of nature, change more quickly than those that are low: though there are exceptions to this rule. The amount of organic change, as Pictet has remarked, does not strictly correspond with the succession of our geological formations; so that between each two consecutive formations, the forms of life have seldom changed in exactly the same degree. Yet if we compare any but the most closely related formations, all the species will be found to have undergone some change. When a species has once disappeared from the face of the earth, we have reason to believe that the same identical form never reappears. The strongest apparent exception to this latter rule, is that of the so-called "colonies" of M. Barrande, which intrude for a period in the midst of an older formation, and then allow the pre-existing fauna to reappear; but Lyell's explanation, namely, that it is a case of temporary migration from a distinct geographical province, seems to me satisfactory.

These several facts accord well with my theory. I believe in no fixed law of development, causing all the inhabitants of a country to change abruptly, or simultaneously, or to an equal degree. The process of modification must be extremely slow. The variability of each species is quite independent of that of all others. Whether such variability be taken advantage of by natural selection, and whether the variations be accumulated to a greater or lesser amount, thus causing a greater or lesser amount of modification in the varying species, depends on many complex contingencies, — on the variability being of a beneficial nature, on the power of intercrossing, on the rate of breeding, on the slowly changing physical conditions of the country, and more especially on the nature of the other inhabitants with which the varying species comes into competition. Hence it is by no means surprising that one species should retain the same identical form much longer than others; or, if changing, that it should change less. We see the same fact in geographical distribution; for instance, in the land-shells and coleopterous insects of Madeira having come to differ considerably from their nearest allies on the continent of Europe, whereas the marine shells and birds have remained unaltered. We can perhaps understand the apparently quicker rate of change in terrestrial and in more highly organised productions compared with marine and lower productions, by the more complex relations of the higher beings to their organic and inorganic conditions of life, as explained in a former chapter. When many of the inhabitants of a country have become modified and improved, we can understand, on the principle of competition, and on that of the many all-important relations of organism to organism, that any form which does not become in some degree modified and improved, will be liable to be exterminated. Hence we can see why all the species in the same region do at last, if we look to wide enough intervals of time, become modified; for those which do not change will become extinct.

※

We can clearly understand why a species when once lost should never reappear, even if the very same conditions of life, organic and inorganic, should recur. For though the offspring of one species might be adapted (and no doubt this has occurred in innumerable instances) to fill the exact place of another species in the economy of nature, and thus supplant it; yet the two forms — the old and the new — would not be identically the same; for both would almost certainly inherit different characters from their distinct progenitors.

※

We have seen in the last chapter that the species of a group sometimes falsely appear to have come in abruptly; and I have attempted to give an explanation of this fact, which if true would have been fatal to my views. But such cases are certainly exceptional; the general rule being a gradual increase in number, till the group reaches its maximum, and then, sooner or later, it gradually decreases. If the number of the species of a genus, or the number of the genera of a family, be represented by a vertical line of varying thickness, crossing the successive geological formations in which the species are found, the line will sometimes falsely appear to begin at its lower end, not in a sharp point, but abruptly; it then gradually thickens upwards, sometimes keeping for a space of equal thickness, and ultimately thins out in the upper beds, marking the decrease and final extinction of the species. This gradual increase in number of the species of a group is strictly conformable with my theory; as the species of the same genus, and the genera of the same family, can increase only slowly and progressively; for the process of modification and the production of a number of allied forms must be slow and gradual, — one species giving rise first to two or three varieties, these being slowly converted into species, which in their turn produce by equally slow steps other species, and so on, like the branching of a great tree from a single stem, till the group becomes large.

On Extinction.— We have as yet spoken only incidentally of the disappearance of species and of groups of species. On the theory of natural selection the extinction of old forms and the production of new and improved forms are intimately connected together. The old notion of all the inhabitants of the earth having been swept away at successive periods by catastrophes, is very generally given up, even by those geologists, as Elie de Beaumont, Murchison, Barrande, &c., whose general views would naturally lead them to this conclusion. On the contrary, we have every reason to believe, from the study of the tertiary formations, that species and groups of species gradually disappear, one after another, first from one spot, then from another, and finally from the world. Both single species and whole groups of species last for very unequal periods; some groups, as we have seen, having endured from the earliest known dawn of life to the present day; some having disappeared before the close of the palaeozoic period. No fixed law seems to determine the length of time during which any single species or any single genus endures. There is reason to believe that the complete extinction of the species of a group is generally a slower process than their production: if the appearance and disappearance of a group of species be represented, as before, by a vertical line of varying thickness, the line is found to taper more gradually at its upper end, which marks the progress of extermination, than at its lower end, which marks the first appearance and increase in numbers of the species. In some cases, however, the extermination of whole groups of beings, as of ammonites towards the close of the secondary period, has been wonderfully sudden.

It is most difficult always to remember that the increase of every living being is constantly being checked by unperceived injurious agencies; and that these same unperceived agencies are amply sufficient to cause rarity, and finally extinction. We see in many cases in the more recent tertiary formations, that rarity precedes extinction; and we know that this has been the progress of events with those animals which have been exterminated, either locally or wholly, through man's agency. I may repeat what I published in 1845, namely, that to admit that species generally become rare before they become extinct —to feel no surprise at the rarity of a species, and yet to marvel greatly when it ceases to exist, is much the same as to admit that sickness in the individual is the forerunner of death — to feel no surprise at sickness, but when the sick man dies, to wonder and to suspect that he died by some unknown deed of violence.

The theory of natural selection is grounded on the belief that each new variety, and ultimately each new species, is produced and maintained by having some advantage over those with which it comes into competition; and the consequent extinction of less-favoured forms almost inevitably follows. It is the same with our domestic productions: when a new and slightly improved variety has been raised, it at first supplants the less improved varieties in the same neighbourhood; when much improved it is transported far and near, like our short-horn cattle, and takes the place of other breeds in other countries. Thus the appearance of new forms and the disappearance of old forms, both natural and artificial, are bound together. In certain flourishing groups, the number of new specific forms which have been produced within a given time is probably greater than that of the old forms which have been exterminated; but we know that the number of species has not gone on indefinitely increasing, at least during the later geological periods, so that looking to later times we may believe that the production of new forms has caused the extinction of about the same number of old forms.

With respect to the apparently sudden extermination of whole families or orders, as of Trilobites at the close of the palaeozoic period and of Ammonites at the close of the secondary period, we must remember what has been already said on the probable wide intervals of time between our consecutive formations; and in these intervals there may have been much slow extermination. Moreover, when by sudden immigration or by unusually rapid development, many species of a new group have taken possession of a new area, they will have exterminated in a correspondingly rapid manner many of the old inhabitants; and the forms which thus yield their places will commonly be allied, for they will partake of some inferiority in common.

Thus, as it seems to me, the manner in which single species and whole groups of species become extinct, accords well with the theory of natural selection. We need not marvel at extinction; if we must marvel, let it be at our presumption in imagining for a moment that we understand the many complex contingencies, on which the existence of each species depends. If we forget for an instant, that each species tends to increase inordinately, and that some check is always in action, yet seldom perceived by us, the whole economy of nature will be utterly obscured. Whenever we can precisely say why this species is more abundant in individuals than that; why this species and not another can be naturalised in a given country; then, and not till then, we may justly feel surprise why we cannot account for the extinction of this particular species or group of species.

On the Affinities of extinct Species to each other, and to living forms.— Let us now look to the mutual affinities of extinct and living species. They all fall into one grand natural system; and this fact is at once explained on the principle of descent. The more ancient any form is, the more, as a general rule, it differs from living forms. But, as Buckland long ago remarked, all fossils can be classed either in still existing groups, or between them. That the extinct forms of life help to fill up the wide intervals between existing genera, families, and orders, cannot be disputed. For if we confine our attention either to the living or to the extinct alone, the series is far less perfect than if we combine both into one general system. With respect to the Vertebrata, whole pages could be filled with striking illustrations from our great palaeontologist, Owen, showing how extinct animals fall in between existing groups. Cuvier ranked the Ruminants and Pachyderms, as the two most distinct orders of mammals; but Owen has discovered so many fossil links, that he has had to alter the whole classification of these two orders; and has placed certain pachyderms in the same sub-order with ruminants: for example, he dissolves by fine gradations the apparently wide difference between the pig and the camel. In regard to the Invertebrata, Barrande, and a higher authority could not be named, asserts that he is every day taught that palaeozoic animals, though belonging to the same orders, families, or genera with those living at the present day, were not at this early epoch limited in such distinct groups as they now are.

Some writers have objected to any extinct species or group of species being considered as intermediate between living species or groups. If by this term it is meant that an extinct form is directly intermediate in all its characters between two living forms, the objection is probably valid. But I apprehend that in a perfectly natural classification many fossil species would have to stand between living species, and some extinct genera between living genera, even between genera belonging to distinct families. The most common case, especially with respect to very distinct groups, such as fish and reptiles, seems to be, that supposing them to be distinguished at the present day from each other by a dozen characters, the ancient members of the same two groups would be distinguished by a somewhat lesser number of characters, so that the two groups, though formerly quite distinct, at that period made some small approach to each other.

On the Succession of the same Types within the same areas, during the later tertiary periods. — Mr. Clift many years ago showed that the fossil mammals from the Australian caves were closely allied to the living marsupials of that continent. In South America, a similar relationship is manifest, even to an uneducated eye, in the gigantic pieces of armour like those of the armadillo, found in several parts of La Plata; and Professor Owen has shown in the most striking manner that most of the fossil mammals, buried there in such numbers, are related to South American types. This relationship is even more clearly seen in the wonderful collection of fossil bones made by MM. Lund and Clausen in the caves of Brazil. I was so much impressed with these facts that I strongly insisted, in 1839 and 1845, on this "law of succession of types," — on "this wonderful relationship in the same continent between the dead and the living." Professor Owen has subsequently extended the same generalisation to the mammals of the Old World. We see the same law in this author's restorations of the extinct and gigantic birds of New Zealand. We see it also in the birds of the caves of Brazil. Mr. Woodward has shown that the same law holds good with sea-shells, but from the wide distribution of most genera of molluscs, it is not well displayed by them. Other cases could be added, as the relation between the extinct and living land-shells of Madeira; and between the extinct and living brackish-water shells of the Aralo-Caspian Sea.

On the theory of descent with modification, the great law of the long enduring, but not immutable, succession of the same types within the same areas, is at once explained; for the inhabitants of each quarter of the world will obviously tend to leave in that quarter, during the next succeeding period of time, closely allied though in some degree modified descendants. If the inhabitants of one continent formerly differed greatly from those of another continent, so will their modified descendants still differ in nearly the same manner and degree. But after very long intervals of time and after great geographical changes, permitting much inter-migration, the feebler will yield to the more dominant forms, and there will be nothing immutable in the laws of past and present distribution.

Summary of the preceding and present Chapters. — I have attempted to show that the geological record is extremely imperfect; that only a small portion of the globe has been geologically explored with care; that only certain classes of organic beings have been largely preserved in a fossil state; that the number both of specimens and of species, preserved in our museums, is absolutely as nothing compared with the incalculable number of generations which must have passed away even during a single formation; that, owing to subsidence being necessary for the accumulation of fossiliferous deposits thick enough to resist future degradation, enormous intervals of time have elapsed between the successive formations; that there has probably been more extinction during the periods of subsidence, and more variation during the periods of elevation, and during the latter the record will have been least perfectly kept; that each single formation has not been continuously deposited; that the duration of each formation is, perhaps, short compared with the average duration of specific forms; that migration has played an important part in the first appearance of new forms in any one area and formation; that widely ranging species are those which have varied most, and have oftenest given rise to new species; and that varieties have at first often been local. All these causes taken conjointly must have tended to make the geological record extremely imperfect, and will to a large extent explain why we do not find interminable varieties, connecting together all the extinct and existing forms of life by the finest graduated steps.

He who rejects these views on the nature of the geological record will rightly reject my whole theory. For he may ask in vain where are the numberless transitional links which must formerly have connected the closely allied or representative species, found in the several stages of the same great formation. He may disbelieve in the enormous intervals of time which have elapsed between our consecutive formations; he may overlook how important a part migration must have played, when the formations of any one great region alone, as that of Europe, are considered; he may urge the apparent, but often falsely apparent, sudden coming in of whole groups of species. He may ask where are the remains of those infinitely numerous organisms which must have existed long before the first bed of the Silurian system was deposited: I can answer this latter question only hypothetically, by saying that as far as we can see, where our oceans now extend they have for an enormous period extended, and where our oscillating continents now stand they have stood ever since the Silurian epoch; but that long before that period, the world may have presented a wholly different aspect; and that the older continents, formed of formations older than any known to us, may now all be in a metamorphosed condition, or may lie buried under the ocean.

Passing from these difficulties, all the other great leading facts in palaeontology seem to me simply to follow on the theory of descent with modification through natural selection. We can thus understand how it is that new species come in slowly and successively; how species of different classes do not necessarily change together, or at the same rate, or in the same degree; yet in the long run that all undergo modification to some extent. The extinction of old forms is the almost inevitable consequence of the production of new forms. We can understand why when a species has once disappeared it never reappears. Groups of species increase in numbers slowly, and endure for unequal periods of time; for the process of modification is necessarily slow, and depends on many complex contingencies. The dominant species of the larger dominant groups tend to leave many modified descendants, and thus new sub-groups and groups are formed. As these are formed, the species of the less vigorous groups, from their inferiority inherited from a common progenitor, tend to become extinct together, and to leave no modified offspring on the face of the earth. But the utter extinction of a whole group of species may often be a very slow process, from the survival of a few descendants, lingering in protected and isolated situations. When a group has once wholly disappeared, it does not reappear; for the link of generation has been broken.

We can understand how the spreading of the dominant forms of life, which are those that oftenest vary, will in the long run tend to people the world with allied, but modified, descendants; and these will generally succeed in taking the places of those groups of species which are their inferiors in the struggle for existence. Hence, after long intervals of time, the productions of the world will appear to have changed simultaneously.

We can understand how it is that all the forms of life, ancient and recent, make together one grand system; for all are connected by generation. We can understand, from the continued tendency to divergence of character, why the more ancient a form is, the more it generally differs from those now living. Why ancient and extinct forms often tend to fill up gaps between existing forms, sometimes blending two groups previously classed as distinct into one; but more commonly only bringing them a little closer together. The more ancient a form is, the more often, apparently, it displays characters in some degree intermediate between groups now distinct; for the more ancient a form is, the more nearly it will be related to, and consequently resemble, the common progenitor of groups, since become widely divergent. Extinct forms are seldom directly intermediate between existing forms; but are intermediate only by a long and circuitous course through many extinct and very different forms. We can clearly see why the organic remains of closely consecutive formations are more closely allied to each other, than are those of remote formations; for the forms are more closely linked together by generation: we can clearly see why the remains of an intermediate formation are intermediate in character.

The inhabitants of each successive period in the world's history have beaten their predecessors in the race for life, and are, in so far, higher in the scale of nature; and this may account for that vague yet ill-defined sentiment, felt by many palaeontologists, that organisation on the whole has progressed. If it should hereafter be proved that ancient animals resemble to a certain extent the embryos of more recent animals of the same class, the fact will be intelligible. The succession of the same types of structure within the same areas during the later geological periods ceases to be mysterious, and is simply explained by inheritance.

If then the geological record be as imperfect as I believe it to be, and it may at least be asserted that the record cannot be proved to be much more perfect, the main objections to the theory of natural selection are greatly diminished or disappear. On the other hand, all the chief laws of palaeontology plainly proclaim, as it seems to me, that species have been produced by ordinary generation: old forms having been supplanted by new and improved forms of life, produced by the laws of variation still acting round us, and preserved by Natural Selection.

Despite the ecological similarities between the Old World and the New, their respective faunas are quite distinct, as are faunas residing closer together where geographic barriers inhibit their communication. On the other hand, related species of the same genus are on occasion found greatly separated. The puzzle Darwin poses here is how in each of these cases the progenitor species could have surmounted the obstacles to their dispersal from common centers of origin. He speculates first on a global topology quite unlike that of the present, where rising and falling land masses could have created dispersal routes from the centers of origin. (Now, of course, we invoke continental drift as our explanation, recognizing the earth's crust to consist of slowly recycling plates that once brought present land masses into approximation and then eventually apart again.) And finally, he offers here a charming account of his observations and experiments on "accidental" means of transport, involving floating seeds and their conveyance by birds.

CHAPTER XI

Geographical Distribution

In considering the distribution of organic beings over the face of the globe, the first great fact which strikes us is, that neither the similarity nor the dissimilarity of the inhabitants of various regions can be accounted for by their climatal and other physical conditions. Of late, almost every author who has studied the subject has come to this conclusion. The case of America alone would almost suffice to prove its truth: for if we exclude the northern parts where the circumpolar land is almost continuous, all authors agree that one of the most fundamental divisions in geographical distribution is that between the New and Old Worlds; yet if we travel over the vast American continent, from the central parts of the United States to its extreme southern point, we meet with the most diversified conditions; the most humid districts, arid deserts, lofty mountains, grassy plains, forests, marshes, lakes, and great rivers, under almost every temperature. There is hardly a climate or condition in the Old World which cannot be paralleled in the New — at least as closely as the same species generally require; for it is a most rare case to find a group of organisms confined to any small spot, having conditions peculiar in only a slight degree; for instance, small areas in the Old World could be pointed out hotter than any in the New World, yet these are not inhabited by a peculiar fauna or flora. Notwithstanding this parallelism in the conditions of the Old and New Worlds, how widely different are their living productions!

A second great fact which strikes us in our general review is, that barriers of any kind, or obstacles to free migration, are related in a close and important manner to the differences between the productions of various regions. We see this in the great difference of nearly all the terrestrial productions of the New and Old Worlds, excepting in the northern parts, where the land almost joins, and where, under a slightly different climate, there might have been free migration for the northern temperate forms, as there now is for the strictly arctic productions. We see the same fact in the great difference between the inhabitants of Australia, Africa, and South America under the same latitude: for these countries are almost as much isolated from each other as is possible. On each continent, also, we see the same fact; for on the opposite sides of lofty and continuous mountain-ranges, and of great deserts, and sometimes even of large rivers, we find different productions; though as mountain-chains, deserts, &c., are not as impassable, or likely to have endured so long as the oceans separating continents, the differences are very inferior in degree to those characteristic of distinct continents.

A third great fact, partly included in the foregoing statements, is the affinity of the productions of the same continent or sea, though the species themselves are distinct at different points and stations. It is a law of the widest generality, and every continent offers innumerable instances. Nevertheless the naturalist in travelling, for instance, from north to south never fails to be struck by the manner in which successive groups of beings, specifically distinct, yet clearly related, replace each other. He hears from closely allied, yet distinct kinds of birds, notes nearly similar, and sees their nests similarly constructed, but not quite alike, with eggs coloured in nearly the same manner. The plains near the Straits of Magellan are inhabited by one species of Rhea (American ostrich), and northward the plains of La Plata by another species of the same genus; and not by a true ostrich or emeu, like those found in Africa and Australia under the same latitude. On these same plains of La Plata, we see the agouti and bizcacha, animals having nearly the same habits as our hares and rabbits and belonging to the same order of Rodents, but they plainly display an American type of structure. We ascend the lofty peaks of the Cordillera and we find an alpine species of bizcacha; we look to the waters, and we do not find the beaver or musk-rat, but the coypu and capybara, rodents of the American type. Innumerable other instances could be given. If we look to the islands off the American shore, however much they may differ in geological structure, the inhabitants, though they may be all peculiar species, are essentially American. We may look back to past ages, as shown in the last chapter, and we find American types then prevalent on the American continent and in the American seas. We see in these facts some deep organic bond, prevailing throughout space and time, over the same areas of land and water, and independent of their physical conditions. The naturalist must feel little curiosity, who is not led to inquire what this bond is.

This bond, on my theory, is simply inheritance, that cause which alone, as far as we positively know, produces organisms quite like, or, as we see in the case of varieties nearly like each other. The dissimilarity of the inhabitants of different regions may be attributed to modification through natural selection, and in a quite subordinate degree to the direct influence of different physical conditions. The degree of dissimilarity will depend on the migration of the more dominant forms of life from one region into another having been effected with more or less ease, at periods more or less remote; — on the nature and number of the former immigrants; — and on their action and reaction, in their mutual struggles for life; — the relation of organism to organism being, as I have already often remarked, the most important of all relations. Thus the high importance of barriers comes into play by checking migration; as does time for the slow process of modification through natural selection. Widely-ranging species, abounding in individuals, which have already triumphed over many competitors in their own widely-extended homes will have the best chance of seizing on new places, when they spread into new countries. In their new homes they will be exposed to new conditions, and will frequently undergo further modification and improvement; and thus they will become still further victorious, and will produce groups of modified descendants. On this principle of inheritance with modification, we can understand how it is that sections of genera, whole genera, and even families are confined to the same areas, as is so commonly and notoriously the case.

I believe, as was remarked in the last chapter, in no law of necessary development. As the variability of each species is an independent property, and will be taken advantage of by natural selection, only so far as it profits the individual in its complex struggle for life, so the degree of modification in different species will be no uniform quantity. If, for instance, a number of species, which stand in direct competition with each other, migrate in a body into a new and afterwards isolated country, they will be little liable to modification; for neither migration nor isolation in themselves can do anything. These principles come into play only by bringing organisms into new relations with each other, and in a lesser degree with the surrounding physical conditions. As we have seen in the last chapter that some forms have retained nearly the same character from an enormously remote geological period, so certain species have migrated over vast spaces, and have not become greatly modified.

On these views, it is obvious, that the several species of the same genus, though inhabiting the most distant quarters of the world, must originally have proceeded from the same source, as they have descended from the same progenitor. In the case of those species, which have undergone during whole geological periods but little modification, there is not much difficulty in believing that they may have migrated from the same region; for during the vast geographical and climatal changes which will have supervened since ancient times, almost any amount of migration is possible. But in many other cases, in which we have reason to believe that the species of a genus have been produced within comparatively recent times, there is great difficulty on this head. It is also obvious that the individuals of the same species, though now inhabiting distant and isolated regions, must have proceeded from one spot, where their parents were first produced: for, as explained in the last chapter, it is incredible that individuals identically the same should ever have been produced through natural selection from parents specifically distinct.

Means of Dispersal.— Sir C. Lyell and other authors have ably treated this subject. I can give here only the briefest abstract of the more important facts. Change of climate must have had a powerful influence on migration: a region when its climate was different may have been a high road for migration, but now be impassable; I shall, however, presently have to discuss this branch of the subject in some detail. Changes of level in the land must also have been highly influential: a narrow isthmus now separates two marine faunas; submerge it, or let it formerly have been submerged, and the two faunas will now blend or may formerly have blended: where the sea now extends, land may at a former period have connected islands or possibly even continents together, and thus have allowed terrestrial productions to pass from one to the other. No geologist will dispute that great mutations of level have occurred within the period of existing organisms. Edward Forbes insisted that all the islands in the Atlantic must recently have been connected with Europe or Africa, and Europe likewise with America. Other authors have thus hypothetically bridged over every ocean, and have united almost every island to some mainland. If indeed the arguments used by Forbes are to be trusted, it must be admitted that scarcely a single island exists which has not recently been united to some continent. This view cuts the Gordian knot of the dispersal of the same species to the most distant points, and removes many a difficulty: but to the best of my judgment we are not authorized in admitting such enormous geographical changes within the period of existing species. It seems to me that we have abundant evidence of great oscillations of level in our continents; but not of such vast changes in their position and extension, as to have united them within the recent period to each other and to the several intervening oceanic islands. I freely admit the former existence of many islands now buried beneath the sea, which may have served as halting places for plants and for many animals during their migration. In the coral-producing oceans such sunken islands are now marked, as I believe, by rings of coral or atolls standing over them. Whenever it is fully admitted, as I believe it will some day be, that each species has proceeded from a single birthplace, and when in the course of time we know something definite about the means of distribution, we shall be enabled to speculate with security on the former extension of the land. But I do not believe that it will ever be proved that within the recent period continents which are now quite separate, have been continuously, or almost continuously, united with each other, and with the many existing oceanic islands. Several facts in distribution, — such as the great difference in the marine faunas on the opposite sides of almost every continent, — the close relation of the tertiary inhabitants of several lands and even seas to their present inhabitants, — a certain degree of relation (as we shall hereafter see) between the distribution of mammals and the depth of the sea, — these and other such facts seem to me opposed to the admission of such prodigious geographical revolutions within the recent period, as are necessitated on the view advanced by Forbes and admitted by his many followers. The nature and relative proportions of the inhabitants of oceanic islands likewise seem to me opposed to the belief of their former continuity with continents. Nor does their almost universally volcanic composition favour the admission that they are the wrecks of sunken continents; — if they had originally existed as mountain-ranges on the land, some at least of the islands would have been formed, like other mountain-summits, of granite, metamorphic schists, old fossiliferous or other such rocks, instead of consisting of mere piles of volcanic matter.

I must now say a few words on what are called accidental means, but which more properly might be called occasional means of distribution. I shall here confine myself to plants. In botanical works, this or that plant is stated to be ill adapted for wide dissemination; but for transport across the sea, the greater or less facilities may be said to be almost wholly unknown. Until I tried, with Mr. Berkeley's aid, a few experiments, it was not even known how far seeds could resist the injurious action of sea-water. To my surprise I found that out of 87 kinds, 64 germinated after an immersion of 28 days, and a few survived an immersion of 137 days. For convenience sake I chiefly tried small seeds, without the capsule or fruit; and as all of these sank in a few days, they could not be floated across wide spaces of the sea, whether or not they were injured by the salt-water. Afterwards I tried some larger fruits, capsules, &c., and some of these floated for a long time. It is well known what a difference there is in the buoyancy of green and seasoned timber; and it occurred to me that floods might wash down plants or branches, and that these might be dried on the banks, and then by a fresh rise in the stream be washed into the sea. Hence I was led to dry stems and branches of 94 plants with ripe fruit, and to place them on sea water. The majority sank quickly, but some which whilst green floated for a very short time, when dried floated much longer; for instance, ripe hazel-nuts sank immediately, but when dried, they floated for 90 days and afterwards when planted they germinated; an asparagus plant with ripe berries floated for 23 days, when dried it floated for 85 days, and the seeds afterwards germinated: the ripe seeds of Helosciadium sank in two days, when dried they floated for above 90 days, and afterwards germinated. Altogether out of the 94 dried plants, 18 floated for above 28 days, and some of the 18 floated for a very much longer period. So that as $^{64}/_{87}$ seeds germinated after an immersion of 28 days; and as $^{18}/_{94}$ plants with ripe fruit (but not all the same species as in the foregoing experiment) floated, after being dried, for above 28 days, as far as we may infer anything from these scanty facts, we may conclude that the seeds of $^{14}/_{100}$ plants of any country might be floated by sea-currents during 28 days, and would retain their power of germination. In Johnston's Physical Atlas, the average rate of the several Atlantic currents is 33 miles per diem (some currents running at the rate of 60 miles per diem); on this average, the seeds of $^{14}/_{100}$ plants belonging to one country might be floated across 924 miles of sea to another country; and when stranded, if blown to a favourable spot by an inland gale, they would germinate.

Living birds can hardly fail to be highly effective agents in the transportation of seeds. I could give many facts showing how frequently birds of many kinds are blown by gales to vast distances across the ocean. We may I think safely assume that under such circumstances their rate of flight would often be 35 miles an hour; and some authors have given a far higher estimate. I have never seen an instance of nutritious seeds passing through the intestines of a bird; but hard seeds of fruit will pass uninjured through even the digestive organs of a turkey. In the course of two months, I picked up in my garden 12 kinds of seeds, out of the excrement of small birds, and these seemed perfect, and some of them, which I tried, germinated. But the following fact is more important: the crops of birds do not secrete gastric juice, and do not in the least injure, as I know by trial, the germination of seeds; now after a bird has found and devoured a large supply of food, it is positively asserted that all the grains do not pass into the gizzard for 12 or even 18 hours. A bird in this interval might easily be blown to the distance of 500 miles, and hawks are known to look out for tired birds, and the contents of their torn crops might thus readily get scattered. Mr. Brent informs me that a friend of his had to give up flying carrier-pigeons from France to England, as the hawks on the English coast destroyed so many on their arrival. Some hawks and owls bolt their prey whole, and after an interval of from twelve to twenty hours, disgorge pellets, which, as I know from experiments made in the Zoological Gardens, include seeds capable of germination. Some seeds of the oat, wheat, millet, canary, hemp, clover, and beet germinated after having been from twelve to twenty-one hours in the stomachs of different birds of prey; and two seeds of beet grew after having been thus retained for two days and fourteen hours. Fresh-water fish, I find, eat seeds of many land and water plants: fish are frequently devoured by birds, and thus the seeds might be transported from place to place. I forced many kinds of seeds into the stomachs of dead fish, and then gave their bodies to fishing-eagles, storks, and pelicans; these birds after an interval of many hours, either rejected the seeds in pellets or passed them in their excrement; and several of these seeds retained their power of germination. Certain seeds, however, were always killed by this process.

Although the beaks and feet of birds are generally quite clean, I can show that earth sometimes adheres to them: in one instance I removed twenty-two grains of dry argillaceous earth from one foot of a partridge, and in this earth there was a pebble quite as large as the seed of a vetch. Thus seeds might occasionally be transported to great distances; for many facts could be given showing that soil almost everywhere is charged with seeds. Reflect for a moment on the millions of quails which annually cross the Mediterranean; and can we doubt that the earth adhering to their feet would sometimes include a few minute seeds? But I shall presently have to recur to this subject.

Considering that the several above means of transport, and that several other means, which without doubt remain to be discovered, have been in action year after year, for centuries and tens of thousands of years, it would I think be a marvellous fact if many plants had not thus become widely transported. These means of transport are sometimes called accidental, but this is not strictly correct: the currents of the sea are not accidental, nor is the direction of prevalent gales of wind. It should be observed that scarcely any means of transport would carry seeds for very great distances; for seeds do not retain their vitality when exposed for a great length of time to the action of sea-water; nor could they be long carried in the crops or intestines of birds. These means, however, would suffice for occasional transport across tracts of sea some hundred miles in breadth, or from island to island, or from a continent to a neighbouring island, but not from one distant continent to another. The floras of distant continents would not by such means become mingled in any great degree; but would remain as distinct as we now see them to be. The currents, from their course, would never bring seeds from North America to Britain, though they might and do bring seeds from the West Indies to our western shores, where, if not killed by so long an immersion in salt-water, they could not endure our climate. Almost every year, one or two land-birds are blown across the whole Atlantic Ocean, from North America to the western shores of Ireland and England; but seeds could be transported by these wanderers only by one means, namely, in dirt sticking to their feet, which is in itself a rare accident. Even in this case, how small would the chance be of a seed falling on favourable soil, and coming to maturity! But it would be a great error to argue that because a well-stocked island, like Great Britain, has not, as far as is known (and it would be very difficult to prove this), received within the last few centuries, through occasional means of transport, immigrants from Europe or any other continent, that a poorly-stocked island, though standing more remote from the mainland, would not receive colonists by similar means. I do not doubt that out of twenty seeds or animals transported to an island, even if far less well-stocked than Britain, scarcely more than one would be so well fitted to its new home, as to become naturalised. But this, as it seems to me, is no valid argument against what would be effected by occasional means of transport, during the long lapse of geological time, whilst an island was being upheaved and formed, and before it had become fully stocked with inhabitants. On almost bare land, with few or no destructive insects or birds living there, nearly every seed, which chanced to arrive, would be sure to germinate and survive.

In this second of two chapters on biogeography, Darwin enlarges the theme established in the last and examines the distribution of island life for its evidence of evolution. Here his observations of earlier years, especially in the Galapagos, serve him amply. Why are oceanic islands devoid of the living forms which lack the means of transport there? Why are less mobile species so often unique to an island? Why have island faunas their greatest affinity with those of the closest mainland, and why are the separate species inhabiting nearby islands often so closely related? The answers weigh heavily against ecclesiastical interpretations. Hardly a more compelling case for evolution could be built on circumstantial evidence than Darwin gives here.

CHAPTER XII

Geographical Distribution — continued

WE NOW come to the last of the three classes of facts, which I have selected as presenting the greatest amount of difficulty, on the view that all the individuals both of the same and of allied species have descended from a single parent; and therefore have all proceeded from a common birthplace, notwithstanding that in the course of time they have come to inhabit distant points of the globe. I have already stated that I cannot honestly admit Forbes's view on continental extensions, which, if legitimately followed out, would lead to the belief that within the recent period all existing islands have been nearly or quite joined to some continent. This view would remove many difficulties, but it would not, I think, explain all the facts in regard to insular productions. In the following remarks I shall not confine myself to the mere question of dispersal; but shall consider some other facts, which bear on the truth of the two theories of independent creation and of descent with modification.

Although in oceanic islands the number of kinds of inhabitants is scanty, the proportion of endemic species (*i.e.* those found nowhere else in the world) is often extremely large. If we compare, for instance, the number of the endemic landshells in Madeira, or of the endemic birds in the Galapagos Archipelago, with the number found on any continent, and then compare the area of the islands with that of the continent, we shall see that this is true. This fact might have been expected on my theory, for, as already explained, species occasionally arriving after long intervals in a new and isolated district, and having to compete with new associates, will be eminently liable to modification, and will often produce groups of modified descendants. But it by no means follows, that, because in an island nearly all the species of one class are peculiar, those of another class, or of another section of the same class, are peculiar; and this difference seems to depend on the species which do not become modified having immigrated with facility and in a body, so that their mutual relations have not been much disturbed. Thus in the Galapagos Islands nearly every land-bird, but only two out of the eleven marine birds, are peculiar; and it is obvious that marine birds could arrive at these islands more easily than land-birds.

Oceanic islands are sometimes deficient in certain classes, and their places are apparently occupied by the other inhabitants; in the Galapagos Islands reptiles, and in New Zealand gigantic wingless birds, take the place of mammals. In the plants of the Galapagos Islands, Dr. Hooker has shown that the proportional numbers of the different orders are very different from what they are elsewhere. Such cases are generally accounted for by the physical conditions of the islands; but this explanation seems to me not a little doubtful. Facility of immigration, I believe, has been at least as important as the nature of the conditions.

With respect to the absence of whole orders on oceanic islands, Bory St. Vincent long ago remarked that Batrachians (frogs, toads, newts) have never been found on any of the many islands with which the great oceans are studded. I have taken pains to verify this assertion, and I have found it strictly true. I have, however, been assured that a frog exists on the mountains of the great island of New Zealand; but I suspect that this exception (if the information be correct) may be explained through glacial agency. This general absence of frogs, toads, and newts on so many oceanic islands cannot be accounted for by their physical conditions; indeed it seems that islands are peculiarly well fitted for these animals; for frogs have been introduced into Madeira, the Azores, and Mauritius, and have multiplied so as to become a nuisance. But as these animals and their spawn are known to be immediately killed by sea-water, on my view we can see that there would be great difficulty in their transportal across the sea, and therefore why they do not exist on any oceanic island. But why, on the theory of creation, they should not have been created there, it would be very difficult to explain.

Mammals offer another and similar case. I have carefully searched the oldest voyages, but have not finished my search; as yet I have not found a single instance, free from doubt, of a terrestrial mammal (excluding domesticated animals kept by the natives) inhabiting an island situated above 300 miles from a continent or great continental island; and many islands situated at a much less distance are equally barren. The Falkland Islands, which are inhabited by a wolf-like fox, come nearest to an exception; but this group cannot be considered as oceanic, as it lies on a bank connected with the mainland; moreover, icebergs formerly brought boulders to its western shores, and they may have formerly transported foxes, as so frequently now happens in the arctic regions. Yet it cannot be said that small islands will not support small mammals, for they occur in many parts of the world on very small islands, if close to a continent; and hardly an island can be named on which our smaller quadrupeds have not become naturalised and greatly multiplied. It cannot be said, on the ordinary view of creation, that there has not been time for the creation of mammals; many volcanic islands are sufficiently ancient, as shown by the stupendous degradation which they have suffered and by their tertiary strata: there has also been time for the production of endemic species belonging to other classes; and on continents it is thought that mammals appear and disappear at a quicker rate than other and lower animals. Though terrestrial mammals do not occur on oceanic islands, aërial mammals do occur on almost every island. New Zealand possesses two bats found nowhere else in the world: Norfolk Island, the Viti Archipelago, the Bonin Islands, the Caroline and Marianne Archipelagoes, and Mauritius, all possess their peculiar bats. Why, it may be asked, has the supposed creative force produced bats and no other mammals on remote islands? On my view this question can easily be answered; for no terrestrial mammal can be transported across a wide space of sea, but bats can fly across.

All the foregoing remarks on the inhabitants of oceanic islands, — namely, the scarcity of kinds — the richness in endemic forms in particular classes or sections of classes, — the absence of whole groups, as of batrachians, and of terrestrial mammals notwithstanding the presence of aërial bats, — the singular proportions of certain orders of plants, — herbaceous forms having been developed into trees, &c., — seem to me to accord better with the view of occasional means of transport having been largely efficient in the long course of time, than with the view of all our oceanic islands having been formerly connected by continuous land with the nearest continent; for on this latter view the migration would probably have been more complete; and if modification be admitted, all the forms of life would have been more equally modified, in accordance with the paramount importance of the relation of organism to organism.

The most striking and important fact for us in regard to the inhabitants of islands is their affinity to those of the nearest mainland, without being actually the same species. Numerous instances could be given of this fact. I will give only one, that of the Galapagos Archipelago, situated under the equator, between 500 and 600 miles from the shores of South America. Here almost every product of the land and water bears the unmistakeable stamp of the American continent. There are twenty-six land birds, and twenty-five of these are ranked by Mr. Gould as distinct species, supposed to have been created here; yet the close affinity of most of these birds to American species in every character, in their habits, gestures, and tones of voice, was manifest. So it is with the other animals, and with nearly all the plants, as shown by Dr. Hooker in his admirable memoir on the Flora of this archipelago. The naturalist, looking at the inhabitants of these volcanic islands in the Pacific, distant several hundred miles from the continent, yet feels that he is standing on American land. Why should this be so? why should the species which are supposed to have been created in the Galapagos Archipelago, and nowhere else, bear so plain a stamp of affinity to those created in America? There is nothing in the conditions of life, in the geological nature of the islands, in their height or climate, or in the proportions in which the several classes are associated together, which resembles closely the conditions of the South American coast: in fact there is a considerable dissimilarity in all these respects. On the other hand, there is a considerable degree of resemblance in the volcanic nature of the soil, in climate, height, and size of the islands, between the Galapagos and Cape de Verde Archipelagos: but what an entire and absolute difference in their inhabitants! The inhabitants of the Cape de Verde Islands are related to those of Africa, like those of the Galapagos to America. I believe this grand fact can receive no sort of explanation on the ordinary view of independent creation; whereas on the view here maintained, it is obvious that the Galapagos Islands would be likely to receive colonists, whether by occasional means of transport or by formerly continuous land, from America; and the Cape de Verde Islands from Africa; and that such colonists would be liable to modification; — the principle of inheritance still betraying their original birthplace.

The principle which determines the general character of the fauna and flora of oceanic islands, namely that the inhabitants, when not identically the same, yet are plainly related to the inhabitants of that region whence colonists could most readily have been derived, — the colonists having been subsequently modified and better fitted to their new homes, — is of the widest application throughout nature. We see this on every mountain, in every lake and marsh. For Alpine species, excepting in so far as the same forms, chiefly of plants, have spread widely throughout the world during the recent Glacial epoch, are related to those of the surrounding lowlands; — thus we have in South America, Alpine humming-birds, Alpine rodents, Alpine plants, &c., all of strictly American forms, and it is obvious that a mountain, as it became slowly upheaved, would naturally be colonised from the surrounding lowlands. So it is with the inhabitants of lakes and marshes, excepting in so far as great facility of transport has given the same general forms to the whole world. We see this same principle in the blind animals inhabiting the caves of America and Europe. Other analogous facts could be given. And it will, I believe, be universally found to be true, that wherever in two regions, let them be ever so distant, many closely allied or representative species occur, there will likewise be found some identical species, showing, in accordance with the foregoing view, that at some former period there has been intercommunication or migration between the two regions. And wherever many closely-allied species occur, there will be found many forms which some naturalists rank as distinct species, and some as varieties; these doubtful forms showing us the steps in the process of modification.

In considering the wide distribution of certain genera, we should bear in mind that some are extremely ancient, and must have branched off from a common parent at a remote epoch; so that in such cases there will have been ample time for great climatal and geographical changes and for accidents of transport; and consequently for the migration of some of the species into all quarters of the world, where they may have become slightly modified in relation to their new conditions. There is, also, some reason to believe from geological evidence that organisms low in the scale within each great class generally change at a slower rate than the higher forms; and consequently the lower forms will have had a better chance of ranging widely and of still retaining the same specific character. This fact, together with the seeds and eggs of many low forms being very minute and better fitted for distant transportation, probably accounts for a law which has long been observed, and which has lately been admirably discussed by Alph. de Candolle in regard to plants, namely, that the lower any group of organisms is, the more widely it is apt to range.

The relations just discussed, — namely, low and slowly-changing organisms ranging more widely than the high, — some of the species of widely-ranging genera themselves ranging widely, — such facts, as alpine, lacustrine, and marsh productions being related (with the exceptions before specified) to those on the surrounding low lands and dry lands, though these stations are so different — the very close relation of the distinct species which inhabit the islets of the same archipelago, — and especially the striking relation of the inhabitants of each whole archipelago or island to those of the nearest mainland, — are, I think, utterly inexplicable on the ordinary view of the independent creation of each species, but are explicable on the view of colonisation from the nearest and readiest source, together with the subsequent modification and better adaptation of the colonists to their new homes.

※

This chapter is not quite the miscellaneous assortment that its title suggests; its unity is given by the Natural System, or systematics — the hierarchical classification of living forms. This all-encompassing system of classification, devised by Linnaeus in the previous century to reveal the plan of the Creator, here Darwin transforms into the roadmap for evolutionary descent. Although comparative anatomy had long been the handmaiden of systematics, only recently had much importance been given rudimentary structures, largely under the influence of Richard Owen, Britain's leading anatomist. Emphasizing the correspondences between seemingly disparate vertebrate skeletal structures (e.g., the forelimbs of bats, moles, and whales), Owen had relied upon rudimentary structures to identify homologies that bore on his theory of archetypes, ideal schemata giving the structural basis of diverse living forms. Here we see Darwin appropriating Owen's evidence in his own cause, giving it his evolutionary interpretation of descent with modification. Embryology had also played a growing, albeit more modest, role in systematics, and Darwin is similarly prepared to enlist its findings to his advantage.

CHAPTER XIII

Mutual Affinities of Organic Beings: Morphology: Embryology: Rudimentary Organs

FROM the first dawn of life, all organic beings are found to resemble each other in descending degrees, so that they can be classed in groups under groups. This classification is evidently not arbitrary like the grouping of the stars in constellations. The existence of groups would have been of simple signification, if one group had been exclusively fitted to inhabit the land, and another the water; one to feed on flesh, another on vegetable matter, and so on; but the case is widely different in nature; for it is notorious how commonly members of even the same sub-group have different habits. In our second and fourth chapters, on Variation and on Natural Selection, I have attempted to show that it is the widely ranging, the much diffused and common, that is the dominant species belonging to the larger genera, which vary most. The varieties, or incipient species, thus produced ultimately become converted, as I believe, into new and distinct species; and these, on the principle of inheritance, tend to produce other new and dominant species. Consequently the groups which are now large, and which generally include many dominant species, tend to go on increasing indefinitely in size. I further attempted to show that from the varying descendants of each species trying to occupy as many and as different places as possible in the economy of nature, there is a constant tendency in their characters to diverge. This conclusion was supported by looking at the great diversity of the forms of life which, in any small area, come into the closest competition, and by looking to certain facts in naturalisation.

Naturalists try to arrange the species, genera, and families in each class, on what is called the Natural System. But what is meant by this system? Some authors look at it merely as a scheme for arranging together those living objects which are most alike, and for separating those which are most unlike; or as an artificial means for enunciating, as briefly as possible, general propositions, — that is, by one sentence to give the characters common, for instance, to all mammals, by another those common to all carnivora, by another those common to the dog-genus, and then by adding a single sentence, a full description is given of each kind of dog. The ingenuity and utility of this system are indisputable. But many naturalists think that something more is meant by the Natural System; they believe that it reveals the plan of the Creator; but unless it be specified whether order in time or space, or what else is meant by the plan of the Creator, it seems to me that nothing is thus added to our knowledge. Such expressions as that famous one of Linnaeus, and which we often meet with in a more or less concealed form, that the characters do not make the genus, but that the genus gives the characters, seem to imply that something more is included in our classification, than mere resemblance. I believe that something more is included; and that propinquity of descent, — the only known cause of the similarity of organic beings, — is the bond, hidden as it is by various degrees of modification, which is partially revealed to us by our classifications.

Let us now consider the rules followed in classification, and the difficulties which are encountered on the view that classification either gives some unknown plan of creation, or is simply a scheme for enunciating general propositions and of placing together the forms most like each other. It might have been thought (and was in ancient times thought) that those parts of the structure which determined the habits of life, and the general place of each being in the economy of nature, would be of very high importance in classification. Nothing can be more false. No one regards the external similarity of a mouse to a shrew, of a dugong to a whale, of a whale to a fish, as of any importance. These resemblances, though so intimately connected with the whole life of the being, are ranked as merely "adaptive or analogical characters;" but to the consideration of these resemblances we shall have to recur. It may even be given as a general rule, that the less any part of the organisation is concerned with special habits, the more important it becomes for classification. As an instance: Owen, in speaking of the dugong, says, "The generative organs being those which are most remotely related to the habits and food of an animal, I have always regarded as affording very clear indications of its true affinities. We are least likely in the modifications of these organs to mistake a merely adaptive for an essential character." So with plants, how remarkable it is that the organs of vegetation, on which their whole life depends, are of little signification, excepting in the first main divisions; whereas the organs of reproduction, with their product the seed, are of paramount importance!

Again, no one will say that rudimentary or atrophied organs are of high physiological or vital importance; yet, undoubtedly, organs in this condition are often of high value in classification. No one will dispute that the rudimentary teeth in the upper jaws of young ruminants, and certain rudimentary bones of the leg, are highly serviceable in exhibiting the close affinity between Ruminants and Pachyderms. Robert Brown has strongly insisted on the fact that the rudimentary florets are of the highest importance in the classification of the Grasses.

Numerous instances could be given of characters derived from parts which must be considered of very trifling physiological importance, but which are universally admitted as highly serviceable in the definition of whole groups. For instance, whether or not there is an open passage from the nostrils to the mouth, the only character, according to Owen, which absolutely distinguishes fishes and reptiles — the inflection of the angle of the jaws in Marsupials — the manner in which the wings of insects are folded — mere colour in certain Algae — mere pubescence on parts of the flower in grasses — the nature of the dermal covering, as hair or feathers, in the Vertebrata. If the Ornithorhynchus had been covered with feathers instead of hair, this external and trifling character would, I think, have been considered by naturalists as important an aid in determining the degree of affinity of this strange creature to birds and reptiles, as an approach in structure in any one internal and important organ.

We can see why characters derived from the embryo should be of equal importance with those derived from the adult, for our classifications of course include all ages of each species. But it is by no means obvious, on the ordinary view, why the structure of the embryo should be more important for this purpose than that of the adult, which alone plays its full part in the economy of nature. Yet it has been strongly urged by those great naturalists, Milne Edwards and Agassiz, that embryonic characters are the most important of any in the classification of animals; and this doctrine has very generally been admitted as true. The same fact holds good with flowering plants, of which the two main divisions have been founded on characters derived from the embryo, — on the number and position of the embryonic leaves or cotyledons, and on the mode of development of the plumule and radicle. In our discussion on embryology, we shall see why such characters are so valuable, on the view of classification tacitly including the idea of descent.

All the foregoing rules and aids and difficulties in classification are explained, if I do not greatly deceive myself, on the view that the natural system is founded on descent with modification; that the characters which naturalists consider as showing true affinity between any two or more species are those which have been inherited from a common parent, and, in so far, all true classification is genealogical; that community of descent is the hidden bond which naturalists have been unconsciously seeking, and not some unknown plan of creation, or the enunciation of general propositions, and the mere putting together and separating objects more or less alike.

Embryology. — It has already been casually remarked that certain organs in the individual, which when mature become widely different and serve for different purposes, are in the embryo exactly alike. The embryos, also, of distinct animals within the same class are often strikingly similar: a better proof of this cannot be given, than a circumstance mentioned by Agassiz, namely, that having forgotten to ticket the embryo of some vertebrate animal, he cannot now tell whether it be that of a mammal, bird, or reptile. The vermiform larvae of moths, flies, beetles, &c., resemble each other much more closely than do the mature insects; but in the case of larvae, the embryos are active, and have been adapted for special lines of life. A trace of the law of embryonic resemblance, sometimes lasts till a rather late age: thus birds of the same genus, and of closely allied genera, often resemble each other in their first and second plumage; as we see in the spotted feathers in the thrush group. In the cat tribe, most of the species are striped or spotted in lines; and stripes can be plainly distinguished in the whelp of the lion. We occasionally though rarely see something of this kind in plants: thus the embryonic leaves of the ulex or furze, and the first leaves of the phyllodineous acaceas, are pinnate or divided like the ordinary leaves of the leguminosae.

The points of structure, in which the embryos of widely different animals of the same class resemble each other often have no direct relation to their conditions of existence. We cannot, for instance, suppose that in the embryos of the vertebrata the peculiar loop-like course of the arteries near the branchial slits are related to similar conditions, — in the young mammal which is nourished in the womb of its mother, in the egg of the bird which is hatched in a nest, and in the spawn of a frog under water. We have no more reason to believe in such a relation, than we have to believe that the same bones in the hand of a man, wing of a bat, and fin of a porpoise, are related to similar conditions of life. No one will suppóse that the stripes on the whelp of a lion, or the spots on the young blackbird, are of any use to these animals, or are related to the conditions to which they are exposed.

The case, however, is different when an animal during any part of its embryonic career is active, and has to provide for itself. The period of activity may come on earlier or later in life; but whenever it comes on, the adaptation of the larva to its conditions of life is just as perfect and as beautiful as in the adult animal. From such special adaptations, the similarity of the larvae or active embryos of allied animals is sometimes much obscured; and cases could be given of the larvae of two species, or of two groups of species, differing quite as much, or even more, from each other than do their adult parents. In most cases, however, the larvae, though active, still obey more or less closely the law of common embryonic resemblance. Cirripedes afford a good instance of this: even the illustrious Cuvier did not perceive that a barnacle was, as it certainly is, a crustacean; but a glance at the larva shows this to be the case in an unmistakeable manner. So again the two main divisions of cirripedes, the pedunculated and sessile, which differ widely in external appearance, have larvae in all their several stages barely distinguishable.

As all the organic beings, extinct and recent, which have ever lived on this earth have to be classed together, and as all have been connected by the finest gradations, the best, or indeed, if our collections were nearly perfect, the only possible arrangement, would be genealogical. Descent being on my view the hidden bond of connexion which naturalists have been seeking. under the term of the natural system. On this view we can understand how it is that in the eyes of most naturalists, the structure of the embryo is even more important for classification than that of the adult. For the embryo is the animal in its less modified state; and in so far it reveals the structure of its progenitor. In two groups of animal, however much they may at present differ from each other in structure and habits, if they pass through the same or similar embryonic stages, we may feel assured that they have both descended from the same or nearly similar parents, and are therefore in that degree closely related. Thus, community in embryonic structure reveals community of descent. It will reveal this community of descent, however much the structure of the adult may have been modified and obscured; we have seen, for instance, that cirripedes can at once be recognised by their larvae as belonging to the great class of crustaceans. As the embryonic state of each species and group of species partially shows us the structure of their less modified ancient progenitors, we can clearly see why ancient and extinct forms of life should resemble the embryos of their descendants, — our existing species.

Thus, as it seems to me, the leading facts in embryology, which are second in importance to none in natural history, are explained on the principle of slight modifications not appearing, in the many descendants from some one ancient progenitor, at a very early period in the life of each, though perhaps caused at the earliest, and being inherited at a corresponding not early period. Embryology rises greatly in interest, when we thus look at the embryo as a picture, more or less obscured, of the common parent-form of each great class of animals.

In this chapter Darwin expresses his confidence that "young and rising naturalists" will share the encompassing unity of his vision. His trust was to be rewarded, of course, but a long campaign lay ahead. The only reference to human evolution in Origin is given here, and that by a single, enigmatic sentence: "Light will be thrown on the origin of man and his history."

CHAPTER XIV

Recapitulation and Conclusion

As this whole volume is one long argument, it may be convenient to the reader to have the leading facts and inferences briefly recapitulated.

That many and grave objections may be advanced against the theory of descent with modification through natural selection, I do not deny. I have endeavoured to give to them their full force. Nothing at first can appear more difficult to believe than that the more complex organs and instincts should have been perfected, not by means superior to, though analogous with, human reason, but by the accumulation of innumerable slight variations, each good for the individual possessor. Nevertheless, this difficulty, though appearing to our imagination insuperably great, cannot be considered real if we admit the following propositions, namely, — that gradations in the perfection of any organ or instinct, which we may consider, either do now exist or could have existed, each good of its kind, — that all organs and instincts are, in ever so slight a degree, variable, — and, lastly, that there is a struggle for existence leading to the preservation of each profitable deviation of structure or instinct. The truth of these propositions cannot, I think, be disputed.

[For brevity, the recapitulation itself has been omitted.]

I have now recapitulated the chief facts and considerations which have thoroughly convinced me that species have changed, and are still slowly changing by the preservation and accumulation of successive slight favourable variations. Why, it may be asked, have all the most eminent living naturalists and geologists rejected this view of the mutability of species? It cannot be asserted that organic beings in a state of nature are subject to no variation; it cannot be proved that the amount of variation in the course of long ages is a limited quantity; no clear distinction has been, or can be, drawn between species and well-marked varieties. It cannot be maintained that species when intercrossed are invariably sterile, and varieties invariably fertile; or that sterility is a special endowment and sign of creation. The belief that species were immutable productions was almost unavoidable as long as the history of the world was thought to be of short duration; and now that we have acquired some idea of the lapse of time, we are too apt to assume, without proof, that the geological record is so perfect that it would have afforded us plain evidence of the mutation of species, if they had undergone mutation.

But the chief cause of our natural unwillingness to admit that one species has given birth to other and distinct species, is that we are always slow in admitting any great change of which we do not see the intermediate steps. The difficulty is the same as that felt by so many geologists, when Lyell first insisted that long lines of inland cliffs had been formed, and great valleys excavated, by the slow action of the coast-waves. The mind cannot possibly grasp the full meaning of the term of a hundred million years; it cannot add up and perceive the full effects of many slight variations, accumulated during an almost infinite number of generations.

Although I am fully convinced of the truth of the views given in this volume under the form of an abstract, I by no means expect to convince experienced naturalists whose minds are stocked with a multitude of facts all viewed, during a long course of years, from a point of view directly opposite to mine. It is so easy to hide our ignorance under such expressions as the "plan of creation," "unity of design," &c., and to think that we give an explanation when we only restate a fact. Any one whose disposition leads him to attach more weight to unexplained difficulties than to the explanation of a certain number of facts will certainly reject my theory. A few naturalists, endowed with much flexibility of mind, and who have already begun to doubt on the immutability of species, may be influenced by this volume; but I look with confidence to the future, to young and rising naturalists, who will be able to view both sides of the question with impartiality. Whoever is led to believe that species are mutable will do good service by conscientiously expressing his conviction; for only thus can the load of prejudice by which this subject is overwhelmed be removed.

Several eminent naturalists have of late published their belief that a multitude of reputed species in each genus are not real species; but that other species are real, that is, have been independently created. This seems to me a strange conclusion to arrive at. They admit that a multitude of forms, which till lately they themselves thought were special creations, and which are still thus looked at by the majority of naturalists, and which consequently have every external characteristic feature of true species, — they admit that these have been produced by variation, but they refuse to extend the same view to other and very slightly different forms. Nevertheless they do not pretend that they can define, or even conjecture, which are the created forms of life, and which are those produced by secondary laws. They admit variation as a *vera causa* in one case, they arbitrarily reject it in another, without assigning any distinction in the two cases. The day will come when this will be given as a curious illustration of the blindness of preconceived opinion.

These authors seem no more startled at a miraculous act of creation than at an ordinary birth. But do they really believe that at innumerable periods in the earth's history certain elemental atoms have been commanded suddenly to flash into living tissues? Do they believe that at each supposed act of creation one individual or many were produced? Were all the infinitely numerous kinds of animals and plants created as eggs or seed, or as full grown? and in the case of mammals, were they created bearing the false marks of nourishment from the mother's womb? Although naturalists very properly demand a full explanation of every difficulty from those who believe in the mutability of species, on their own side they ignore the whole subject of the first appearance of species in what they consider reverent silence.

It may be asked how far I extend the doctrine of the modification of species. The question is difficult to answer, because the more distinct the forms are which we may consider, by so much the arguments fall away in force. But some arguments of the greatest weight extend very far. All the members of whole classes can be connected together by chains of affinities, and all can be classified on the same principle, in groups subordinate to groups. Fossil remains sometimes tend to fill up very wide intervals between existing orders. Organs in a rudimentary condition plainly show that an early progenitor had the organ in a fully developed state; and this in some instances necessarily implies an enormous amount of modification in the descendants. Throughout whole classes various structures are formed on the same pattern, and at an embryonic age the species closely resemble each other. Therefore I cannot doubt that the theory of descent with modification embraces all the members of the same class. I believe that animals have descended from at most only four or five progenitors, and plants from an equal or lesser number.

Analogy would lead me one step further, namely, to the belief that all animals and plants have descended from some one prototype. But analogy may be a deceitful guide. Nevertheless all living things have much in common, in their chemical composition, their germinal vesicles, their cellular structure, and their laws of growth and reproduction. We see this even in so trifling a circumstance as that the same poison often similarly affects plants and animals; or that the poison secreted by the gall-fly produces monstrous growths on the wild rose or oak-tree. Therefore I should infer from analogy that probably all the organic beings which have ever lived on this earth have descended from some one primordial form, into which life was first breathed.

When the views entertained in this volume on the origin of species, or when analogous views are generally admitted, we can dimly foresee that there will be a considerable revolution in natural history. Systematists will be able to pursue their labours as at present; but they will not be incessantly haunted by the shadowy doubt whether this or that form be in essence a species. This I feel sure, and I speak after experience, will be no slight relief. The endless disputes whether or not some fifty species of British brambles are true species will cease. Systematists will have only to decide (not that this will be easy) whether any form be sufficiently constant and distinct from other forms, to be capable of definition; and if definable, whether the differences be sufficiently important to deserve a specific name. This latter point will become a far more essential consideration than it is at present; for differences, however slight, between any two forms, if not blended by intermediate gradations, are looked at by most naturalists as sufficient to raise both forms to the rank of species. Hereafter we shall be compelled to acknowledge that the only distinction between species and well-marked varieties is, that the latter are known, or believed, to be connected at the present day by intermediate gradations, whereas species were formerly thus connected. Hence, without quite rejecting the consideration of the present existence of intermediate gradations between any two forms, we shall be led to weigh more carefully and to value higher the actual amount of difference between them. It is quite possible that forms now generally acknowledged to be merely varieties may hereafter be thought worthy of specific names, as with the primrose and cowslip; and in this case scientific and common language will come into accordance. In short, we shall have to treat species in the same manner as those naturalists treat genera, who admit that genera are merely artificial combinations made for convenience. This may not be a cheering prospect; but we shall at least be freed from the vain search for the undiscovered and undiscoverable essence of the term species.

The other and more general departments of natural history will rise greatly in interest. The terms used by naturalists of affinity, relationship, community of type, paternity, morphology, adaptive characters, rudimentary and aborted organs, &c., will cease to be metaphorical, and will have a plain signification. When we no longer look at an organic being as a savage looks at a ship, as at something wholly beyond his comprehension; when we regard every production of nature as one which has had a history; when we contemplate every complex structure and instinct as the summing up of many contrivances, each useful to the possessor, nearly in the same way as when we look at any great mechanical invention as the summing up of the labour, the experience, the reason, and even the blunders of numerous workmen; when we thus view each organic being, how far more interesting, I speak from experience, will the study of natural history become!

A grand and almost untrodden field of inquiry will be opened, on the causes and laws of variation, on correlation of growth, on the effects of use and disuse, on the direct action of external conditions, and so forth. The study of domestic productions will rise immensely in value. A new variety raised by man will be a far more important and interesting subject for study than one more species added to the infinitude of already recorded species. Our classifications will come to be, as far as they can be so made, genealogies; and will then truly give what may be called the plan of creation. The rules for classifying will no doubt become simpler when we have a definite object in view. We possess no pedigrees or armorial bearings; and we have to discover and trace the many diverging lines of descent in our natural genealogies, by characters of any kind which have long been inherited. Rudimentary organs will speak infallibly with respect to the nature of long-lost structures. Species and groups of species, which are called aberrant, and which may fancifully be called living fossils, will aid us in forming a picture of the ancient forms of life. Embryology will reveal to us the structure, in some degree obscured, of the prototypes of each great class.

When we can feel assured that all the individuals of the same species, and all the closely allied species of most genera, have within a not very remote period descended from one parent, and have migrated from some one birthplace; and when we better know the many means of migration, then, by the light which geology now throws, and will continue to throw, on former changes of climate and of the level of the land, we shall surely be enabled to trace in an admirable manner the former migrations of the inhabitants of the whole world. Even at present, by comparing the differences of the inhabitants of the sea on the opposite sides of a continent, and the nature of the various inhabitants of that continent in relation to their apparent means of immigration, some light can be thrown on ancient geography.

The noble science of Geology loses glory from the extreme imperfection of the record. The crust of the earth with its embedded remains must not be looked at as a well-filled museum, but as a poor collection made at hazard and at rare intervals. The accumulation of each great fossiliferous formation will be recognised as having depended on an unusual concurrence of circumstances, and the blank intervals between the successive stages as having been of vast duration. But we shall be able to gauge with some security the duration of these intervals by a comparison of the preceding and succeeding organic forms. We must be cautious in attempting to correlate as strictly contemporaneous two formations, which include few identical species, by the general succession of their forms of life. As species are produced and exterminated by slowly-acting and still existing causes, and not by miraculous acts of creation and by catastrophes; and as the most important of all causes of organic change is one which is almost independent of altered and perhaps suddenly altered physical conditions, namely, the mutual relation of organism to organism, — the improvement of one being entailing the improvement or the extermination of others; it follows, that the amount of organic change in the fossils of consecutive formations probably serves as a fair measure of the lapse of actual time. A number of species, however, keeping in a body might remain for a long period unchanged, whilst within this same period, several of these species, by migrating into new countries and coming into competition with foreign associates, might become modified; so that we must not overrate the accuracy of organic change as a measure of time. During early periods of the earth's history, when the forms of life were probably fewer and simpler, the rate of change was probably slower; and at the first dawn of life, when very few forms of the simplest structure existed, the rate of change may have been slow in an extreme degree. The whole history of the world, as at present known, although of a length quite incomprehensible by us, will hereafter be recognised as a mere fragment of time, compared with the ages which have elapsed since the first creature, the progenitor of innumerable extinct and living descendants, was created.

In the distant future I see open fields for far more important researches. Psychology will be based on a new foundation, that of the necessary acquirement of each mental power and capacity by gradation. Light will be thrown on the origin of man and his history.

Authors of the highest eminence seem to be fully satisfied with the view that each species has been independently created. To my mind it accords better with what we know of the laws impressed on matter by the Creator, that the production and extinction of the past and present inhabitants of the world should have been due to secondary causes, like those determining the birth and death of the individual. When I view all beings not as special creations, but as the lineal descendants of some few beings which lived long before the first bed of the Silurian system was deposited, they seem to me to become ennobled. Judging from the past, we may safely infer that not one living species will transmit its unaltered likeness to a distant futurity. And of the species now living very few will transmit progeny of any kind to a far distant futurity; for the manner in which all organic beings are grouped shows that the greater number of species of each genus, and all the species of many genera, have left no descendants, but have become utterly extinct. We can so far take a prophetic glance into futurity as to foretel that it will be the common and widely-spread species, belonging to the larger and dominant groups, which will ultimately prevail and procreate new and dominant species. As all the living forms of life are the lineal descendants of those which lived long before the Silurian epoch, we may feel certain that the ordinary succession by generation has never once been broken, and that no cataclysm has desolated the whole world. Hence we may look with some confidence to a secure future of equally inappreciable length. And as natural selection works solely by and for the good of each being, all corporeal and mental endowments will tend to progress towards perfection.

It is interesting to contemplate an entangled bank, clothed with many plants of many kinds, with birds singing on the bushes, with various insects flitting about, and with worms crawling through the damp earth, and to reflect that these elaborately constructed forms, so different from each other, and dependent on each other in so complex a manner, have all been produced by laws acting around us. These laws, taken in the largest sense, being Growth with Reproduction; Inheritance which is almost implied by reproduction; Variability from the indirect and direct action of the external conditions of life, and from use and disuse; a Ratio of Increase so high as to lead to a Struggle for Life, and as a consequence to Natural Selection, entailing Divergence of Character and the Extinction of less-improved forms. Thus, from the war of nature, from famine and death, the most exalted object which we are capable of conceiving, namely, the production of the higher animals, directly follows. There is grandeur in this view of life, with its several powers, having been originally breathed into a few forms or into one; and that, whilst this planet has gone cycling on according to the fixed law of gravity, from so simple a beginning endless forms most beautiful and most wonderful have been, and are being, evolved.

5

The Theory Defended

So IMMEDIATE was the response to *Origin* that, by the close of 1860, one commentator saw fit to claim that

> No scientific work that has been published within this century has excited so much general curiosity as the treatise of Mr. Darwin. It has for a time divided the scientific world with two great contending sections. A Darwinite and an anti-Darwinite are now the badges of opposed scientific parties.[1]

To understand the initial reactions of the scientific community, it helps to know something of the disputants involved. At the core of this Darwinite party, all with close personal attachments to Darwin himself, were Sir Charles Lyell, Joseph Hooker, and Thomas Huxley. The oldest of the three, Lyell, was easily the most prominent, both within the scientific community and, as a companion of Prince Albert, without. Although not then an evolutionist himself, Lyell had encouraged Darwin to write up his theory in the first place, had along with Hooker communicated it to the Linnean Society, and finally had vouchsafed its commercial and scientific merit to his own publisher. Certainly among these Darwinites, Lyell's prestigious endorsements could benefit Darwin the most; yet he proved to be the most recalcitrant convert of the three, finding continual difficulty in going more than part way with Darwin. Their correspondence reveals Lyell promising regularly to declare wholeheartedly in favor of evolution, only then to vacillate and withdraw his pledge.[2] In the *Antiquity of Man* (1863), for example, having already assured Darwin he was prepared to "go the whole orang," he went no further than to admit to a saltationist theory of evolution, with natural selection playing a subordinate role to guidance by "preconcerted arrangement," an ambiguity that seemed to imply divine intervention.[3] Not until 1868, in the tenth edition of *Principles of Geology*, did Lyell announce himself more or less squarely in support, and even then he hinted at a Providential design undergirding the evolutionary process. Still, however much Lyell's equivocation irritated Darwin, it went largely unnoticed by others, who perceived little distance separating their positions. To outward appearances at least, Lyell was a Darwinite.

Joseph Hooker, the eminent botanist, was a far readier recruit. Like Darwin, Hooker in his earlier years also served as a ship's naturalist, voyaging to New Zealand, Australia, and the Antarctic aboard the H.M.S. *Erebus* to catalogue the flora there. During Darwin's early London years they became fast friends, and Hooker was the first to share Darwin's evolutionary views. Although not a convert at once, by the date of *Origin*'s publication, Hooker had already revealed his allegiance, having produced the first of his many works to demonstrate the applicability of natural selection to the interpretation of botanical diversity. Of all the Darwinites, Hooker held a position on evolution that most closely resembled Darwin's in its details. When he succeeded his father as director of Kew Gardens, he occupied a post that proved an influential platform in promoting Darwinism.

Easily the most visible standard-bearer of the Darwinites was Thomas Huxley, the self-proclaimed "bulldog of Darwinism." He had launched his scientific career much as Hooker had done somewhat earlier, as a surgeon-naturalist aboard the H.M.S. *Rattlesnake* during its South Pacific voyages. Following his return in 1850 he soon distinguished himself as a brilliant comparative anatomist. The ambitious Huxley gained further prominence by a series of aggressive attacks on Richard Owen, whose theory of archetypal design he dismissed as an out-moded Platonism (see p. 235 and Ruse[4] for further details). Like Hooker, Huxley did not embrace Darwinism immediately; introduced to his friend's theory some years earlier, not until the publication of *Origin* did he announce his conversion. Once proselytized, though, he took up Darwin's case with unmatched zeal and considerable flair. Flaying its adversaries with mighty rhetorical weaponry, he directed the richest of his polemic gifts against the Darwinites' numerous clerical critics. Still, Huxley himself had certain doubts about evolution according to Darwin. His most significant reservation was with the claim that new species arose perforce by gradual and insensible change. "You have loaded yourself with an unnecessary difficulty in adopting *natura non facit saltum* so unreservedly," he wrote to Darwin after first reading *Origin*.[5] In Huxley's view, evolution could proceed by leaps and often did.

Many others were identifiably in Darwin's camp, beyond his innermost circle of close friends. In America, for example, the botanist Asa Gray was the spearhead of the Darwinite campaign. Sometimes called the "American Huxley," Gray vigorously weighed in against the antievolutionary pronouncements of his renowned Harvard colleague, the zoologist Louis Agassiz. Gray's defense of evolution is especially interesting for the heterodoxy of his own views on the matter. As a devout Christian, Gray found a home for natural theology even in the Darwinian scheme by proposing the evolutionary process itself to be divinely guided. In this he was far more explicit than Lyell. Thus, to rephrase Darwinism in accord with Paley's argument from design (p. 53), there cannot be natural selection without a Natural Selector. Eventually Gray did relinquish his claim that selection itself was divinely governed, but only to recast Providence in its custodial role as the Originator of Variation. Gray thus managed to bracket the materialism of Darwinism with an overarching teleology of divine intelligence. Indeed, during the first decades after the publication of *Origin* this became a widely favored interpretation. (The interested reader will find in Ghiselin[6] and Gillespie[7] more extensive discussions of evolution in light of theological concerns.)

Alfred Wallace's part in this contest deserves comment as well. As the cofounder of evolution by natural selection and as Darwin's dedicated supporter, his placement in the Darwinite constellation is irrefutable. Yet seen in more sociological terms, Wallace played only an adjunct role in shaping the course of debate after *Origin*. Partly this was because he remained in the Far East until 1862, by which time the lines had long been drawn and the debate well under way. But this is not the whole story, for his stature in the scientific community seems grossly incommensurate with the seminal quality of his contribution. With neither university training nor affiliations, he was regarded as hardly more than a scholarly amateur, a view which his own deference probably helped to reinforce. As an outsider, Wallace had little opportunity to create a following (even supposing that he were so disposed); Darwin, by contrast, recruited supporters to his banner assiduously. Even the influential evolutionary work of the entymologist Henry Bates, Wallace's old comrade with whom he had earlier explored the Amazon, seems to have been more strongly marked by Darwin's stamp.

If Wallace's standing among the Darwinites was chiefly honorific, at least so far as their politics went, his evolutionary views touched a number of their various ideological bases. On one hand, there was no more vigorous advocate than he of the creative power of natural selection, including even Darwin himself. Just as he disputed Darwin's dismissal of natural selection to explain reproductive isolation between species, so was he equally critical of Darwin's ready tendency to explain the evolution of most secondary sex characteristics by sexual selection. Where Darwin saw vivid male plumage among birds the result of their competition for mates, for example, Wallace saw

instead the drabness of females as their protection from pred-
ators. But like Lyell and Gray, on the other hand, Wallace was
unwilling to abandon evolution to material forces altogether.
Thus he eventually called upon miraculous intervention to explain what for him was the special case of the human mind
and spirit. It is perhaps worth mention in this connection that
later in his life, when he redirected his enthusiasm from evolution to various nonscientific pursuits, prominent among them
were phrenology and spiritualism.

Thus we see that among the Darwinites, evolutionary opinions wandered well beyond the perimeter of Darwin's own ideas.
With saltation advocated by Huxley and Lyell and creationist
accounts of one stripe or another advanced by Lyell, Gray, and
Wallace, the defenders of Darwinism were deployed far from
their center. It was their good fortune that the opponents they
faced were equally disarrayed in their opinions.

The opening salvoes in the conflict took the form of reviews
of *Origin* in popular periodicals, published anonymously as was
then the custom. Within the month following *Origin*'s publication, for example, Hooker wrote an appreciative review for
Gardener's Chronicle, while Huxley provided several, including an important one for the *Times*. This latter assignment he
contrived to obtain from a friend, a baffled staff writer on that
newspaper who unsuspectingly had solicited Huxley's assistance on the project. But here the anti-Darwinites were not to
be outdone. Shortly thereafter, the Cambridge geologist Adam
Sedgwick (who had long before introduced Darwin to field-
work as his expedition leader in North Wales) issued a thun-
derous broadside against *Origin* in the *Spectator*. His review
found the case for evolution wanting on geological grounds,
on its methodology, and, no less, on moral grounds. It was the
righteous abusiveness of Sedgwick's attack that led Darwin to
identify correctly its author, who months before had personally
rebuked him by letter for his heresies:

There is a moral or metaphysical part of nature as well as a physical. A man who denies this is deep in the mire of folly. 'Tis the crown and glory of organic science that it *does* through *final cause*, link material and moral. . . . You have ignored this link; and, if I do not mistake your meaning, you have done your best in one or two cases to break it. Were it possible (which thank God it is not) to break it, humanity, in my mind, would suffer a damage that might brutalize it, and sink the human race into a lower grade of degradation than any into which it has fallen since its written records tell us of its history.[8]

Like many of his colleagues at Cambridge and Oxford, Sedgwick was also an Anglican cleric. The tender accord between science and revealed religion having been breached, his stand on the side of orthodoxy is therefore unremarkable. Yet most of his copartisans (and here I refer only to those with one foot planted in science) took some pains to emphasize in their reviews the proper independence of science from theology, even if only to find in favor of the latter wherever their respective spheres collided. Thus the zoologist Thomas Wollaston could write that, regarding science and theology, "they are best discussed apart, and that neither of them was ever intended to teach us the other," but shortly thereafter comment that "no man who loves truth, in all its phases, for it own sake, will long rest in accepting *as such* a zoological creed which is in direct antagonism with his theological one."[9]

Just as all the Darwinites were not materialists, however, neither were all the anti-Darwinites motivated by their religion. Early in the conflict an especially scathing review of *Origin* appeared in the *Edinburgh Review*, one which gave the usual criticisms of Darwin's theory but at the same time allowed "no sympathy whatever with Biblical objectors to creation by law, or with the sacerdotal revilers of those who would explain such law," probably an oblique reference to Sedgwick.[10] The author was Richard Owen, easily identified by the profusion of adulatory references to the publications and views of the same. Owen, who had long been secretly hatching an evolutionary scheme, took the opportunity of anonymity both to attack Darwin's and to promote his own. The latter he accomplished first by finding redeeming merit in the notorious progressionist teleology first advanced by Chambers in *Vestiges of the Natural History of Creation* (see pp. 52–54) — which in its major conception resembled his own — and then by favorably citing precedents in his own writings. The defense of Chambers required no little hypocrisy on Owen's part, since he had himself earlier excoriated *Vestiges of Creation* anonymously (no doubt chagrined by its premature revelation of his own theory), while privately lending support to its author. Again we find Owen repeating the same sordid ploy, encouraging Darwin privately, assailing him publicly, while claiming his own priority to evolutionary theory (see Himmelfarb[11] for further details).

Owen's actual position on evolution, outlined in the *Edinburgh Review* and then more completely in several works published over the decade, is virtually impossible to ascertain. Certainly Darwin grew weary of the effort, for in the *Historical Sketch* beginning the sixth edition of *Origin* he pithily recounts his attempts:

> When the first edition of this work was published, I was so completely deceived, as were many others, by such expressions as 'the continuous operation of creative power,' that I included Professor Owen with the other paleontologists as being firmly convinced of the immutability of species; but it appears . . . that this was on my part a preposterous error. In the last edition of this work I inferred, and the inference still to me seems perfectly just, . . . that Professor Owen admitted that natural selection may have done something in the formation of a new species; but this it appears . . . is inaccurate and without evidence. I also gave some extracts from correspondence between Professor Owen and the editor of the 'London Review,' from which it appeared manifest to the Editor as well as to myself, that Professor

Owen claimed to have promulgated the theory of natural selection before I had done so; and I expressed my surprise and satisfaction at this announcement; but as far as it is possible to understand certain recently published passages . . . I have either partially or wholly fallen into error. It is consolatory to me that others find Professor Owen's controversial writings as difficult to understand and to reconcile with each other as I do.

I also confess an inability to penetrate very far the mysteries of Owen's thesis, but Michael Ruse[12] has offered as comprehensible an exposition as the modern reader might want.

Forthrightness was not Owen's outstanding habit. Following his denunciation of "sacerdotal revilers," he threw in with Samuel Wilberforce, Bishop of Oxford, to mount an attack on *Origin* from the *Quarterly Review*, a review that surpassed all then extant polemic standards for its exaggerated misrepresentation of Darwin's views. Not only did the Bishop lengthily revile Darwin for the errancy of his theology, but also for the evolutionary sins of his grandfather, Erasmus Darwin. This alliance between cleric and scientist led to the most theatrical moment in the entire contest over Darwinism, the Wilberforce–Huxley debate. The bout was staged at the British Association meeting of 1860, held at Oxford that year. Darwin, whose practice was to avoid such gatherings, was not in attendance, but Hooker and Huxley were. During a session earlier in the week Owen had propounded his thesis that a man's brain is unique by its singular possession of a hippocampus minor, a claim which Huxley contradicted with a pledge to publish a timely rebuttal. This so enraged Owen that he presented his tutorial services to Wilberforce, scheduled to speak at the final Saturday session. On this event Wilberforce launched his assault before a large assemblage of laymen, undergraduates, and academics, delivering much the same scurrilous charges published soon thereafter in the *Quarterly Review*. But Wilberforce strayed from the line of criticism tendered by Owen, and in an infelicitous attempt to ridicule his opposition, he made so grievous a tactical blunder as to assure instantly his celebration by posterity. Turning to Huxley, he inquired whether it was through his grandfather or grandmother that he claimed descent from a monkey. When the time to speak passed to Huxley, he issued first a solemn defense of Darwinism, and only then addressed Wilberforce's slight; posed with the question " 'would I rather have a miserable ape for a grandfather, or a man highly endowed by nature and possessed of great influence, and yet who employs these faculties and that influence for the mere purpose of introducing ridicule into a grave scientific discussion' — I unhesitatingly affirm my preference for the ape."[13] This riposte was sufficient to precipitate an uproarious commotion: undergraduates leaped up and shouted, at least one woman fainted and was carried from the proceedings, while Fitzroy stamped about in the audience, invoking the Bible and waving aloft the object of his concern. The meeting was disbanded after Hooker, its final speaker, affirmed his Darwinian faith. (For more details of the spectacle, see Himmelfarb.[14])

Despite its drama, the Oxford victory was not the decisive engagement in the conflict — no single occasion was pivotal, in fact — but soon thereafter the tide began to run visibly in the Darwinites' favor. Their numbers grew steadily, drawn especially from the ranks of those "young and rising naturalists" whom Darwin had expressly hoped to recruit. In 1864 Darwin was awarded the greatest honor of the Royal Society, the Copley medal (receiving it, as was his custom, in absentia). But his theory was still to face its most formidable opposition.

In 1867 a sharply critical review of the fourth edition of *Origin* appeared in *The North British Review*;[15] its author, Fleeming Jenkin, was a Scottish engineer who had participated with Sir William Thomson (later Lord Kelvin) in the laying of the transatlantic cable. Much of Jenkin's criticism had been already offered by others, yet taken as a whole its effect on Darwin was considerable, for as he wrote to Hooker:

It is only about two years since last edition of *Origin* and I am fairly disgusted to find how much I have to modify, and how much I ought to add; but I have determined not to add much. Fleeming Jenkin has given me much trouble, but has been of more real use to me than any other essay or review.[16]

Jenkin's review merits some close consideration because it exposed the vulnerability of Darwin's theory as it then stood, predicated on an erroneous concept of blending inheritance, and because of the nature of Darwin's eventual response to it.

Citing the physicist Kelvin's estimate of one hundred million years for the duration of life on earth, Jenkin faulted Darwin's geologically derived estimate (see Chapter IX, *Origin*) and the uniformitarian premises upon which it was founded. But the case had been made earlier by Kelvin himself, who based his computations on the rate of global cooling from a molten state; he had, of course, failed to take into account the earth's then unknown radioactivity. Darwin had his own misgivings about the validity of any such estimates — in fact, by the third edition of *Origin* he struck out even his own calculations of the period required for denudation of the Weald — and he resolved to let the future decide the matter. Indeed, there was little else for him to do.

Jenkin also took issue with Darwin's reading of the fossil record. Even granting Darwin's claim for intermittent episodes of preservation, he asked, why did not even those fragments as *were* preserved reveal insensible and gradual changes, as if in transition from one variety to another? Faced often enough with the question, Darwin had already made an important concession here. In the fourth edition (1866), after his earlier explanation for the unlikelihood of finding transitional evolutionary stages in a single geological formation (see p. 174), he added the paragraph:

It is a more important consideration, leading to the same result, as lately insisted on by Dr. Falconer, namely, that the period during which each species underwent modification, though long as measured by years was probably short in comparison with that during which it remained without undergoing any change.

While it would be too much to claim that Darwin intended by this a radical retrenchment from his original position of gradualism, it is nevertheless a significant rescaling of the process of natural selection, shifting it from a more or less continuous phenomenon to an intermittent, albeit still gradual, operation. This same theme was echoed in *The Variation of Animals and Plants under Domestication* (1868), where he wrote:

> It is generally believed, though on this head we have little or no evidence, that new characters become fixed; and after having long remained fixed it seems possible that under new conditions they might again be rendered variable.[17]

Thus, by reformulating his theory to reflect long periods of evolutionary equilibrium occasionally disrupted by change, Darwin largely vitiated this particular criticism by Jenkin. (What I find most interesting about this revision, as I shall discuss in the final chapter, is the manner in which it anticipated a contemporary controversy on the tempo of evolution, the debate on "punctuated equilibrium.")

On the topics of variation and inheritance, however, Jenkin's criticism was more trenchant. Where Darwin had begun his argument for evolution by analogy from domestication, Jenkin beat Darwin with his own stick: a racehorse, however enduring and intensive the breeding of his ancestors, remains nevertheless a horse; the breeder's art may create new varieties, but never new species. Jenkin argued, as had others before him (including not only Wollaston, Agassiz, and Wilberforce among the anti-Darwinites, but Asa Gray as well), that a natural limit on variation restricts the extent to which an organism may depart from others in its species: "Experiments conducted during the longest time at our disposal show no probability of surpassing the limits of the sphere of variation, and why should we concede that a simple extension of time will reverse the rule?"[15] Lacking counter-evidence, all Darwin could do here was restate his belief that there is equally no reason to believe variation to be so constrained. In the fifth edition of *Origin* (1869) he responded:

> Some authors have maintained that the amount of variation in our domestic productions is soon reached, and can never be exceeded. It would be somewhat rash to assert the limit has been attained in any one case; for almost all our animals and plants have been greatly improved within a recent period; and this implies variation. It would be equally rash to assert that characters now increased to their utmost limit could not, after remaining fixed for many centuries, again vary under new conditions of life.

However adroit this parry, Jenkin's attack on the effectiveness of natural selection under blending inheritance was far harder to turn aside. In an especially clear and forceful argument, Jenkin demonstrated that any trait first arising in a population, regardless of how much greater benefit it confers upon its initial bearer, will never become fixed throughout the population because its selective advantage is successively reduced by blending in each following generation. Thus blending inheritance, by swamping any favorable trait which might appear, acts to limit the impact that natural selection can produce. More damaging yet, Jenkin pointed out the two circumstances most favorable to the swamping of a favorable trait — and therefore most constraining on selection — are those in which (1) selection favors an inherited deviation only slightly displaced from the ordinary, since the relative advantage of the trait over all others cannot be very great in the first place, and (2) the trait arises in a large, continuously distributed population, since more individuals exist to receive the favorable variant and dilute its expression. Of course, these were exactly the conditions that Darwin had held to be most propitious for the origin of species by means of natural selection.

But Jenkin was not yet finished. Suppose that natural selection were to act favorably upon "single variations," those rare and more obviously manifest variations also called "sports" or "monstrosities," whose contribution to the evolutionary process Darwin had largely discounted because their bearers tend to be infertile. Suppose further that the descendants of such individuals, however unlikely these be, retain the full expression of the inherited trait. Then, argued Jenkin,

> If . . . the advantage given by the sport is retained by all descendants, independently of what in common speech might be called proportion of blood in their veins directly derived from the first sport, then these descendants will shortly supplant the whole species entirely, after the manner required by Darwin.[15]

If Jenkin first gives the appearance here of offering Darwin an exit from his dilemma, he quickly dispelled it, going on to point out that this "is surely not the Darwinian theory of gradual accumulation of infinitely minute variations of every-day occurrence," lacking completely explanation for either the undiminished expression of the trait through descent or its favorable production in the first place. It is, Jenkin concluded, an account of species creation by unspecified saltatory means.

Although among the Darwinites Huxley and Gray and perhaps even Lyell had claimed a greatly expanded role for larger variations or "sports," Darwin had never responded with much enthusiasm to the idea. The effect of Jenkin's argument was to harden his position on this issue, committing him further to his view that imperceptibly small "individual variations" constitute the basis for evolutionary change. Besides, since the blending theory of inheritance then widely prevailed, the sustained "unimpaired transmission" of a trait that Jenkin hypothecated (and Mendel confirmed) was in any event an unrecognized phenomenon.* Thus Darwin was forced to puzzle out. the conundrum posed by Jenkin's first alternative: How does selection fix a minute variation in the face of blending inheritance?

The solution Darwin arrived at was rooted in his Lamarckian ideas on the sources of variation, most recently expressed in his hypothesis of *pangenesis*. The curious "provisional hypothesis of pangenesis" was the most notable feature of his treatise, *The Variation of Animals and Plants under Domestication* (1868), completed soon after receiving Jenkin's review and after years of intermittent labor. The work as a whole amplified the views — mainly incorrect — on inheritance and variation which he first set forth in *Origin*, and pangenesis was its ill-fated attempt to bolster evolutionary theory with explanation for the principles of heredity upon which it ultimately depended. The kernel of pangenesis involved production by the body's cells of minute particles, *gemmules*, postulated to carry the instructions for cellular recreation in the offspring. Transported by bodily fluids, gemmules were presumed to collect in the germ cells, there to await fertilization when they would mix with their counterparts. The derived properties of pangenesis are the most interesting here, however. As corollaries, Darwin claimed that unusual use of a body part might stimulate excessive production of gemmules, while disuse promoted their deficiency. By further extension, conditions external to the organism might intervene in this sequence, so to shape the pattern of heredity in the offspring. Thus, in positing direct and modifiable connections between the organism's somatic and germ cells, pangenesis gave the theoretical underpinnings for Lamarckian inheritance of acquired traits.

---

* On this subject, Vorzimmer[18] gives intriguing accounts of two "close brushes" Darwin had with particulate or Mendelian inheritance. Vorzimmer has documented that Darwin possessed an abstract of Mendel's classic paper on his breeding experiments, and also a remarkably prescient explanation of particulate inheritance in a letter from Francis Galton. On neither occasion, however, did he grasp the implications for his theory.

The enlistment of these principles in answering Jenkin led Darwin's defense of natural selection in the direction that we might least have expected. Jenkin had challenged him to explain why small variations that might arise would not dissipate by successive outcrossings with nonbearers, instead to accumulate and become fixed. Darwin reasoned that such an outcome ought to follow should the favorable trait arise not merely in a single individual or two, but widely throughout the population, all at once. This, he imagined, would counter the tendency for such a trait to become progressively diminished, because under the circumstances its carriers were likely to mate with others similarly endowed and thereby consolidate the trait among their offspring. It only remained to initiate this phenomenon, and here the Lamarckian mechanism suited Darwin's requirements perfectly, since he held that the "direct effects of environment" as well as the effects of use and disuse of parts would be widely if not uniformly manifest throughout a population, particularly when subject to sustained environmental conditions. Accordingly, in the final two editions of *Origin* (and more clearly in *The Descent of Man*, as we shall see) the inheritance of acquired traits enjoys a much elevated prominence. Compare, for example, this summary passage in the final chapter of the first edition

> The complex and little known laws governing variation are the same, as far as we can see, with the laws which have governed so-called specific forms. In both cases physical conditions seem to have produced *but little direct effect*. . . . In both varieties and species, use and disuse seem to have produced *some effect*. . . .[emphasis added]

with its correspondent in the sixth edition:

> In both cases physical conditions seem to have produced *some direct and definite effect*. . . . With both varieties and species,

use and disuse seem to have produced a *considerable effect.* . . .[emphasis added]

So where the first edition distinctly subordinated the importance of Lamarckian mechanisms to the "indirect effects of environment," in the later renderings of his theory Darwin leaned extensively upon them as the initiators of organic change.* Still, it must be understood that in the confrontation with Jenkin's criticism, Darwin did not retreat from natural selection to become a Lamarckian, as sometimes has been alleged. Of the several mechanisms which he recognized might lead to species modification, he left no doubt that natural selection held pride of place.

Despite Darwin's own perceptions of his theory's integrity, it weathered the engagement with Jenkin only at some cost. Increasing reliance on Lamarckian principles jeopardized at least the primacy given natural selection, if not its necessity also. Not surprisingly, at least one critic, St. George Mivart, saw how Darwin's response to Jenkin had weakened his position. A comparative anatomist, Mivart had been a student of both Owen and Huxley and an early proponent of evolution directed by natural selection. His eventual defection from the Darwinite camp stemmed from his growing inability to reconcile the respective tenets of his dual conversion to Darwinism and to Catholicism. Mivart announced his leave with a strident fusillade of criticism against the materialist foundations of Darwinism, especially as brought to bear upon the subject of mankind. In these misgivings he shared common ground with Wallace. But where Wallace was satisfied to claim special dispensation for man's mental and moral faculties, leaving the balance of evolution to the creative power of natural selection, the scope of Mivart's approach was far less selective. In *The Genesis of Species*[20] and several related articles, Mivart set out to dem-

---

* Vorzimmer[19] gives an excellent account of this transformation in Darwin's thinking, although I find his interpretation of Darwin's final withdrawal slightly overblown.

onstrate the incompleteness of natural selection as an explanation for the whole of organic change. He denied neither the
fact of evolution nor even the process of natural selection —
only the sufficiency of selection to accomplish all that Darwin
claimed for it. Most of the arguments he delivered against
Darwin's view of selection had originated with others, most
notably Jenkin: insensibly small variations are the least heritable; they are too small to provide much benefit — if any — to
their bearers; selection is limited by inherent constraints on
variation; fragments of the fossil record reveal no gradual change;
favorable saltations do occur; and so forth. Moreover, selection
could not account for the appearance of many structures useful to the organism only in their fully evolved form, a change
Mivart supported by numerous examples. Going still further,
Mivart was determined to dislodge Darwin from the high ground
of pangenesis, from which he had previously fended off Jenkin. Thus he assailed pangenesis on the well-chosen grounds
that its poorly specified mechanisms served to obscure more than
they clarified, and that its predicated claims were counterfac-
tual (e.g., the sustained reappearance of the foreskin of Jews
despite millennia of circumcision). On this score Mivart enjoyed influential support; Francis Galton, Darwin's younger
cousin and Britain's foremost student of heredity, had in the
early 1870s performed a series of experiments that largely dis-
credited the mechanisms hypothesized for pangenesis.[21] In the
wake of Galton's and others' findings, the once popular concept of acquired inheritance had begun losing favor.

Notwithstanding the exceedingly hostile tone of Mivart's attack and his egregious distortion of Darwin's actual views, his
battery of arguments against natural selection was then more
astute than any advanced by previous critics. Yet Mivart, as I
have indicated, also gave confirmation both to evolution and
natural selection. This contradiction hints at a paradoxical
quality in the debate during the decade after *Origin*, for there
was little in Mivart's own views to distinguish them from those

of the Darwinite Asa Gray. Like Gray, Mivart saw evolution as a playing out of divinely ordained teleology, where favorable variation arises by providential guidance. Like Mivart, Gray placed a premium on the evolutionary significance of larger variations — and for much the same reasons (although Mivart was somewhat more disposed than Gray to entertain evolution by large saltations). That they stood in opposing ranks had more to do with their personalities than with their science.

Mivart's attack did not go long unanswered. In a puckish gesture, Huxley[22] took Mivart to task with scholastic flourishes for espousing a theology quite dissonant with the true teachings of the Church, Mivart's assertions to the contrary. It was not merely coincidental that Mivart had hoped by *Genesis* to allay papal doubts on evolution. Although Darwin had at first regarded Mivart's criticism benignly enough, he soon received a review of *Genesis* from Chauncey Wright,[23] an American mathematician and philosopher, pointing out Mivart's deliberately specious portrayal of Darwin's arguments. Now persuaded of the iniquity in Mivart's attack, Darwin had Wright's article republished in Britain at his own expense (but only after soliciting unsuccessfully Wallace's aid in rebuttal). If underwriting the costs of his defense was unusual for Darwin, it was altogether characteristic of him not to respond personally, for he had long before determined to avoid occupying himself too closely with the moment's controversy. Instead, as had become his usual practice, he attended to the latest criticisms within the text of his major work in progress.

The sixth edition of *Origin* (1872) furnished the first opportunity for his reply (since *Descent* had already gone to the printer when *Genesis* was published), and he inserted an additional chapter, "Miscellaneous Objections to the Theory of Natural Selection," mainly to address Mivart. In opening his case, Darwin left little doubt that Mivart had acted with bad faith in misconstruing his position:

A distinguished zoologist, Mr. St. George Mivart, has recently collected all objections which have ever been advanced by myself and others against the theory of natural selection, as propounded by Mr. Wallace and myself, and has thus illustrated them with admirable art and force. When thus marshalled, they make a formidable array; and as it forms no part of Mr. Mivart's plan to give the various facts and considerations opposed to his conclusions, no slight effort of reason and memory is left to the reader, who may wish to weigh the evidence on both sides. . . . My judgement may not be trustworthy, but after reading with care Mr. Mivart's book, and comparing each section with what I have said on the same head, I never before felt so strongly convinced of the general truth of the conclusions here arrived at, subject, of course, in so intricate a subject, to much partial error.

What followed was a systematic and cogent rebuttal of Mivart's many examples purporting the existence of structures without adaptive value in their primordial states. Additionally, Darwin defended himself against Mivart's sneering charge that he had abandoned natural selection in the face of the enemy to embrace Lamarckism, rightly claiming that he had given weight to acquired inheritance from the first. What was notably absent from his rejoinder, however, was a counter to the criticism leveled at pangenesis. Since the revision engendered by Jenkin's criticism had elevated the importance of pangenesis to the theory as a whole, Mivart's attack thus had not been entirely without success. In the end, then, the final edition of *Origin* did not quite restore Darwin's theory to the *status quo ante*.

Over the course of its various editions, *Origin* had grown increasingly encumbered by an accumulated weight of emendations, taking on a rather piecemeal appearance in comparison to its original form. *The Descent of Man* (1871) furnished Darwin a fresh tablet, along with the opportunity for redeeming his implied promise to shed light on man's history. Lyell[24] had earlier written on human origins, and although he offered considerable room for speculation on providential guidance, he did allow that certain features may be the result of natural selection. Huxley[25] had also addressed the topic, expectably with less reticence than Lyell, and Wallace had indicated to Darwin that he was waiting in the wings with his interpretation. Thus public sensitivity to the issue of human evolution had greatly abated by the time that Darwin addressed it in *Descent*.

In one sense *Descent* represents a case study in which all Darwin's general evolutionary principles are brought to bear. In another sense, however, the work — and man's part in it — is primarily a vehicle for setting forth his ideas on sexual selection, a theme dominating three-quarters of the book. The distinction may be significant, for viewed the first way we are given the impression that Darwin has come to invest sexual selection with inordinate explanatory power, much at the expense of natural selection. This is, in fact, a common interpretation, supported by Darwin's generally growing reliance upon factors other than natural selection. On the other hand, if we see the work as chiefly Darwin's exploration of the implications of sexual selection for evolution, with human origins assuming an adjunct role, then we cannot expect in his treatment of man an altogether "balanced" or "representative" model of his evolutionary schema. I confess my preference for the latter interpretation, if only because well over half its text has not the least to do with man or primate, being devoted instead to the application of sexual selection to a wide sampling of invertebrate and vertebrate orders. Whatever the case, there can be little question that sexual selection has benefitted by an expanded explanatory role in this enterprise, and that natural selection has incurred a loss thereby.

6

The Descent of Man

In chapter I of The Descent, Darwin established his case for human evolution: man shares with mammals generally and with other primates particularly a constellation of biochemical, physiological, and developmental features. In a similar vein, certain of man's features — e.g., sense of smell, body hair, appendix, and coccyx — are rudimentary with respect to their development in other species, and thus betoken evolutionary reduction from some ancestral state.

I find this chapter, the second in the volume, notable for two quite different reasons. The first is the way in which it so clearly marks the shift in Darwin's thinking since Origin first appeared. Here, with application to man, Darwin uses much the same format that he did in Origin to introduce his theory: man is rich in variety, tends (or once did) to increase in number unless checked by famine and disease, and is subject, therefore, to natural selection. But the emphasis has altered quite perceptibly. Note, for example, the prominence that use and disuse receive as principles of inherited variation, while spontaneous variation (i.e., "indirect effect of external conditions"), a mainstay of the original formulation, receives only a two-sentence paragraph, anomalously under the heading "correlated variation." Natural selection makes a rather subdued appearance here. For example, regarding the anatomical correlates of bipedalism (i.e., characteristics of foot, pelvis, spine, and head), he writes, "It is very difficult to decide how far these correlated modifications are the result of natural selection, and how far of the inherited use of certain parts, or of the action of one part on another." Similarly, the loss of the tail in the apes and man is attributed mainly to "inherited effects of mutilation" which result from chafing during sitting, followed by a lengthy and much-quoted (perhaps first by Mivart) admission that previously too much weight may have been placed upon selection.

The other reason this chapter merits attention is for its remarkable reconstruction of the divergence and initial evolution of the human lineage. Darwin predicates this series of events on a dietary shift, perhaps related to an environmental change, which in turn led our ancestors to abandon their arboreal niche for a terrestrial habitat. Erect bipedalism emerged, founded on an increasing reliance upon tools for defense and hunting, which served (through a feedback mechanism of sophisticated conception) to reinforce the nascent pattern of tool use. Eventually, continued dependence upon tools contributed, via enhanced mental faculties, to neurocranial expansion and, as selection for powerful chewing muscles relaxes, to eventual reduction of facial and dental structures. This scenario for human origins is extraordinary in that Darwin was able to construct it guided by pure reason alone. Prevailing paleontological knowledge was confined to the relatively modern neandertal fossils, removed by millions of years from the stage that he described; neither had archeologists then identified cultural evidence of greater antiquity. Yet this reconstruction served anthropology for an entire century as its favorite origins myth; it suffered relatively few revisions despite the accumulation of substantially greater material evidence of human biocultural evolution. Only during the past decade, under the weight of new findings from East Africa, has this conception been seriously challenged. (The popular summaries by Johanson and Edey[1] and by Leakey and Lewin[2] give the flavor of recent evidence and arguments for various reinterpretations; my own preference is for the former work, which relies upon a more secure fossil chronology.)

Finally, Darwin's citation of differences in cranial capacity *between populations and sexes, indexed to degrees of intellectual development, deserves some brief comment. This misunderstanding was commonplace during the last century and survived well into the present one. The invariably paramount position of Europeans and males on this scale seems to have been as much a product of unmerited racial and sexual chauvinism as it was, more directly, a reflection of inappropriate sampling procedures and a failure to take into account allometric differences — differences resulting from growth in individuals of different size. Moreover, save in cases of gross developmental disruption (e.g., microcephaly), modern inquiry has failed to demonstrate a connection in living man between brain size and intellectual capacity. (Stephen Gould[3] treats engagingly the dismal history of interpretative bias in these matters.)*

CHAPTER II

On the Manner of Development of Man from Some Lower Form

IT IS MANIFEST that man is now subject to much variability. No two individuals of the same race are quite alike. We may compare millions of faces, and each will be distinct. There is an equally great amount of diversity in the proportions and dimensions of the various parts of the body; the length of the legs being one of the most variable points. Although in some quarters of the world an elongated skull, and in other quarters a short skull prevails, yet there is great diversity of shape even within the limits of the same race, as with the aborigines of America and South Australia — the latter a race "probably as pure and homogeneous in blood, customs, and language as any in existence"— and even with the inhabitants of so confined an area as the Sandwich Islands. An eminent dentist assures me that there is nearly as much diversity in the teeth as in the features. The chief arteries so frequently run in abnormal courses, that it has been found useful for surgical purposes to calculate from 1040 corpses how often each course prevails. The muscles are eminently variable: thus those of the foot were found by Prof. Turner not to be strictly alike in any two out of fifty bodies; and in some the deviations were considerable. He adds, that the power of performing the appropriate movements must have been modified in accordance with the several deviations. Mr. J. Wood has recorded the occurrence of 295 muscular variations in thirty-six subjects, and in another set of the same number no less than 558 variations, those occurring on both sides of the body being only reckoned as one. In the last set, not one body out of the thirty-six was "found totally wanting in departures from the standard descriptions of the muscular system given in anatomical text books." A single body presented the extraordinary number of twenty-five distinct abnormalities. The same muscle sometimes varies in many ways: thus Prof. Macalister describes no less than twenty distinct variations in the *palmaris accessorius*.

I have elsewhere so fully discussed the subject of Inheritance, that I need here add hardly anything. A greater number of facts have been collected with respect to the transmission of the most trifling, as well as of the most important characters in man, than in any of the lower animals; though the facts are copious enough with respect to the latter. So in regard to mental qualities, their transmission is manifest in our dogs, horses, and other domestic animals. Besides special tastes and habits, general intelligence, courage, bad and good temper, &c., are certainly transmitted. With man we see similar facts in almost every family; and we now know, through the admirable labours of Mr. Galton, that genius which implies a wonderfully complex combination of high faculties, tends to be inherited; and, on the other hand, it is too certain that insanity and deteriorated mental powers likewise run in families.

With respect to the causes of variability, we are in all cases very ignorant; but we can see that in man as in the lower animals, they stand in some relation to the conditions to which each species has been exposed, during several generations. Domesticated animals vary more than those in a state of nature; and this is apparently due to the diversified and changing nature of the conditions to which they have been subjected. In this respect the different races of man resemble domesticated animals, and so do the individuals of the same race, when inhabiting a very wide area, like that of America. We see the influence of diversified conditions in the more civilised nations; for the members belonging to different grades of rank, and following different occupations, present a greater range of character than do the members of barbarous nations. But the uniformity of savages has often been exaggerated, and in some cases can hardly be said to exist. It is, nevertheless, an error to speak of man, even if we look only to the conditions to which he has been exposed, as "far more domesticated" than any other animal. Some savage races, such as the Australians, are not exposed to more diversified conditions than are many species which have a wide range. In another and much more important respect, man differs widely from any strictly domesticated animal; for his breeding has never long been controlled, either by methodical or unconscious selection. No race or body of men has been so completely subjugated by other men, as that certain individuals should be preserved, and thus unconsciously selected, from somehow excelling in utility to their masters. Nor have certain male and female individuals been intentionally picked out and matched, except in the well-known case of the Prussian grenadiers; and in this case man obeyed, as might have been expected, the law of methodical selection;

for it is asserted that many tall men were reared in the villages inhabited by the grenadiers and their tall wives. In Sparta, also, a form of selection was followed, for it was enacted that all children should be examined shortly after birth; the well-formed and vigorous being preserved, the others left to perish.

If we consider all the races of man as forming a single species, his range is enormous; but some separate races, as the Americans and Polynesians have very wide ranges. It is a well-known law that widely-ranging species are much more variable than species with restricted ranges; and the variability of man may with more truth be compared with that of widely-ranging species, than with that of domesticated animals.

The Direct and Definite Action of Changed Conditions. — This is a most perplexing subject. It cannot be denied that changed conditions produce some, and occasionally a considerable effect, on organisms of all kinds; and it seems at first probable that if sufficient time were allowed this would be the invariable result. But I have failed to obtain clear evidence in favour of this conclusion; and valid reasons may be urged on the other side, at least as far as the innumerable structures are concerned, which are adapted for special ends. There can, however, be no doubt that changed conditions induce an almost indefinite amount of fluctuating variability, by which the whole organisation is rendered in some degree plastic.

In the United States, above 1,000,000 soldiers, who served in the late war, were measured, and the States in which they were born and reared were recorded. From this astonishing number of observations it is proved that local influences of some kind act directly on stature; and we further learn that "the State where the physical growth has in great measure taken place, and the State of birth, which indicates the ancestry, seem to exert a marked influence on the stature." For instance, it is established, "that residence in the Western States, during the years of growth, tends to produce increase of stature." On the other hand, it is certain that with sailors, their life delays growth, as shewn "by the great difference between the statures of soldiers and sailors at the ages of seventeen and eighteen years." Mr. B. A. Gould endeavoured to ascertain the nature of the influences which thus act on stature; but he arrived only at negative results, namely that they did not relate to climate, the elevation of the land, soil, nor even "in any controlling degree" to the abundance or the need of the comforts of life. This latter conclusion is directly opposed to that arrived at by Villermé, from the statistics of the height of the conscripts in different parts of France. When we compare the differences in stature between the Polynesian chiefs and the lower orders within the same islands, or between the inhabitants of the fertile volcanic and low barren coral islands of the same ocean, or again between the Fuegians on the eastern and western shores of their country, where the means of subsistence are very different, it is scarcely possible to avoid the conclusion that better food and greater comfort do influence stature. But the preceding statements shew how difficult it is to arrive at any precise result. Dr. Beddoe has lately proved that, with the inhabitants of Britain, residence in towns and certain occupations have a deteriorating influence on height; and he infers that the result is to a certain extent inherited, as is likewise the case in the United States. Dr. Beddoe further believes that wherever a "race attains its maximum of physical development, it rises highest in energy and moral vigour."

Effects of the increased Use and Disuse of Parts. — It is well known that use strengthens the muscles in the individual, and complete disuse, or the destruction of the proper nerve, weakens them. When the eye is destroyed, the optic nerve often becomes atrophied. When an artery is tied, the lateral channels increase not only in diameter, but in the thickness and strength of their coats. When one kidney ceases to act from disease, the other increases in size, and does double work. Bones increase not only in thickness, but in length, from carrying a greater weight. Different occupations, habitually followed, lead to changed proportions in various parts of the body. Thus it was ascertained by the United States Commission that the legs of the sailors employed in the late war were longer by 0.217 of an inch than those of the soldiers, though the sailors were on an average shorter men; whilst their arms were shorter by 1.09 of an inch, and therefore, out of proportion, shorter in relation to their lesser height. This shortness of the arms is apparently due to their greater use, and is an unexpected result: but sailors chiefly use their arms in pulling, and not in supporting weights. With sailors, the girth of the neck and the depth of the instep are greater, whilst the circumference of the chest, waist, and hips is less, than in soldiers.

Whether the several foregoing modifications would become hereditary, if the same habits of life were followed during many generations, is not known, but it is probable. Rengger attributes the thin legs and thick arms of the Payaguas Indians to successive generations having passed nearly their whole lives in canoes, with their lower extremities motionless. Other writers have come to a similar conclusion in analogous cases. According to Cranz, who lived for a long time with the Esquimaux, "the natives believe that ingenuity and dexterity in seal-catching (their highest art and virtue) is hereditary; there is really something in it, for the son of a celebrated seal-catcher will distinguish himself, though he lost his father in childhood." But in this case it is mental aptitude, quite as much as bodily structure, which appears to be inherited. It is asserted that the hands of English labourers are at birth larger than those of the gentry. From the correlation which exists, at least in some cases, between the development of the extremities and of the jaws, it is possible that in those classes which do not labour much with their hands and feet, the jaws would be reduced in size from this cause. That they are generally smaller in refined and civilized men than in hard-working men or savages, is certain. But with savages, as Mr. Herbert Spencer has remarked, the greater use of the jaws in chewing coarse, uncooked food, would act in a direct manner on the masticatory muscles, and on the bones to which they are attached. In infants, long before birth, the skin on the soles of the feet is thicker than on any other part of the body; and it can hardly be doubted that this is due to the inherited effects of pressure during a long series of generations.

It is familiar to every one that watchmakers and engravers are liable to be short-sighted, whilst men living much out of doors, and especially savages, are generally long-sighted. Short-sight and long-sight certainly tend to be inherited. The inferiority of Europeans, in comparison with savages, in eyesight and in the other senses, is no doubt the accumulated and transmitted effect of lessened use during many generations; for Rengger states that he has repeatedly observed Europeans, who had been brought up and spent their whole lives with the wild Indians, who nevertheless did not equal them in the sharpness of their senses. The same naturalist observes that the cavities in the skull for the reception of the several sense-organs are larger in the American aborigines than in Europeans; and this probably indicates a corresponding difference in the dimensions of the organs themselves. Blumenbach has also remarked on the large size of the nasal cavities in the skulls of the American aborigines, and connects this fact with their remarkably acute power of smell. The Mongolians of the plains of Northern Asia, according to Pallas, have wonderfully perfect senses;

and Prichard believes that the great breadth of their skulls across the zygomas follows from their highly-developed sense-organs.

Although man may not have been much modified during the latter stages of his existence through the increased or decreased use of parts, the facts now given shew that his liability in this respect has not been lost; and we positively know that the same law holds good with the lower animals. Consequently we may infer that when at a remote epoch the progenitors of man were in a transitional state, and were changing from quadrupeds into bipeds, natural selection would probably have been greatly aided by the inherited effects of the increased or diminished use of the different parts of the body.

Arrests of Development. — There is a difference between arrested development and arrested growth, for parts in the former state continue to grow whilst still retaining their early condition. Various monstrosities come under this head; and some, as a cleft palate, are known to be occasionally inherited. It will suffice for our purpose to refer to the arrested brain-development of microcephalous idiots, as described in Vogt's memoir. Their skulls are smaller, and the convolutions of the brain are less complex than in normal men. The frontal sinus, or the projection over the eye-brows, is largely developed, and the jaws are prognathous to an *"effrayant"* degree: so that these idiots somewhat resemble the lower types of mankind. Their intelligence, and most of their mental faculties, are extremely feeble. They cannot acquire the power of speech, and are wholly incapable of prolonged attention, but are much given to imitation. They are strong and remarkably active, continually gambolling and jumping about, and making grimaces. They often ascend stairs on all-fours; and are curiously fond of climbing up furniture or trees. We are thus reminded of the delight shewn by almost all boys in climbing trees; and this again reminds us how lambs and kids, originally alpine animals, delight to frisk on any hillock, however small. Idiots also resemble the lower animals in some other respects; thus several cases are recorded of their carefully smelling every mouthful of food before eating it. One idiot is described as often using his mouth in aid of his hands, whilst hunting for lice. They are often filthy in their habits, and have no sense of decency; and several cases have been published of their bodies being remarkably hairy.

Reversion. — Many of the cases to be here given might have been introduced under the last heading. When a structure is arrested in its development, but still continues growing, until it closely resembles a corresponding structure in some lower and adult member of the same group, it may in one sense be considered as a case of reversion. The lower members in a group give us some idea how the common progenitor was probably constructed; and it is hardly credible that a complex part, arrested at an early phase of embryonic development, should go on growing so as ultimately to perform its proper function, unless it had acquired such power during some earlier state of existence, when the present exceptional or arrested structure was normal. The simple brain of a microcephalous idiot, in as far as it resembles that of an ape, may in this sense be said to offer a case of reversion. There are other cases which come more strictly under our present head of reversion. Certain structures, regularly occurring in the lower members of the group to which man belongs, occasionally make their appearance in him, though not found in the normal human embryo; or, if normally present in the human embryo, they become abnormally developed, although in a manner which is normal in the lower members of the group. These remarks will be rendered clearer by the following illustrations.

In man, the canine teeth are perfectly efficient instruments for mastication. But their true canine character, as Owen remarks, "is indicated by the conical form of the crown, which terminates in an obtuse point, is convex outward and flat or sub-concave within, at the base of which surface there is a feeble prominence. The conical form is best expressed in the Melanian races, especially the Australian. The canine is more deeply implanted, and by a stronger fang than the incisors." Nevertheless, this tooth no longer serves man as a special weapon for tearing his enemies, or prey; it may, therefore, as far as its proper function is concerned, be considered as rudimentary. In every large collection of human skulls some may be found, as Häckel observes, with the canine teeth projecting considerably beyond the others in the same manner as in the anthropomorphous apes, but in a less degree. In these cases, open spaces between the teeth in the one jaw are left for the reception of the canines of the opposite jaw. An inter-space of this kind in a Kaffir skull, figured by Wagner, is surprisingly wide. Considering how few are the ancient skulls which have been examined, compared to recent skulls, it is an interesting fact that in at least three cases the canines project largely; and in the Naulette jaw they are spoken of as enormous.

Of the anthropomorphous apes the males alone have their canines fully developed; but in the female gorilla, and in a less degree in the female orang, these teeth project considerably beyond the others; therefore the fact, of which I have been assured, that women sometimes have considerably projecting canines, is no serious objection to the belief that their occasional great development in man is a case of reversion to an ape-like progenitor. He who rejects with scorn the belief that the shape of his own canines, and their occasional great development in other men, are due to our early forefathers having been provided with these formidable weapons, will probably reveal, by sneering, the line of his descent. For though he no longer intends, nor has the power, to use these teeth as weapons, he will unconsciously retract his "snarling muscles" (thus named by Sir C. Bell), so as to expose them ready for action, like a dog prepared to fight.

Many muscles are occasionally developed in man, which are proper to the Quadrumana or other mammals. Professor Vlacovich examined forty male subjects, and found a muscle, called by him the ischio-pubic, in nineteen of them; in three others there was a ligament which represented this muscle; and in the remaining eighteen no trace of it. In only two out of thirty female subjects was this muscle developed on both sides, but in three others the rudimentary ligament was present. This muscle, therefore, appears to be much more common in the male than in the female sex; and on the belief in the descent of man from some lower form, the fact is intelligible; for it has been detected in several of the lower animals, and in all of these it serves exclusively to aid the male in the act of reproduction.

These various cases of reversion are so closely related to those of rudimentary organs given in the first chapter, that many of them might have been indifferently introduced either there or here. . . . These several reversionary structures, as well as the strictly rudimentary ones, reveal the descent of man from some lower form in an unmistakable manner.

Correlated Variation. — In man, as in the lower animals, many structures are so intimately related, that when one part varies so does another, without our being able, in most cases, to assign any reason. We cannot say whether the one part governs the other, or whether both are governed by some earlier developed part. Various monstrosities, as I. Geoffroy repeatedly insists, are thus intimately connected. Homologous structures are particularly liable to change together, as we see on the opposite sides of the body, and in the upper and lower extremities. Meckel long ago remarked, that when the muscles of the arm depart from their proper type, they almost always imitate those of the leg; and so, conversely, with the muscles of the legs. The organs of sight and hearing, the teeth and hair, the colour of the skin and of the hair, colour and constitution, are more or less correlated. Professor Schaaffhausen first drew attention to the relation apparently existing between a muscular frame and the strongly-pronounced supra-orbital ridges, which are so characteristic of the lower races of man.

Besides the variations which can be grouped with more or less probability under the foregoing heads, there is a large class of variations which may be provisionally called spontaneous, for to our ignorance they appear to arise without any exciting cause. It can, however, be shewn that such variations, whether consisting of slight individual differences, or of strongly-marked and abrupt deviations of structure, depend much more on the constitution of the organism than on the nature of the conditions to which it has been subjected.

Rate of Increase. — Civilised populations have been known under favourable conditions, as in the United States, to double their numbers in twenty-five years; and, according to a calculation, by Euler, this might occur in a little over twelve years. At the former rate, the present population of the United States (thirty millions), would in 657 years cover the whole terraqueous globe so thickly, that four men would have to stand on each square yard of surface. The primary or fundamental check to the continued increase of man is the difficulty of gaining subsistence, and of living in comfort. We may infer that this is the case from what we see, for instance, in the United States, where subsistence is easy, and there is plenty of room. If such means were suddenly doubled in Great Britain, our number would be quickly doubled. With civilised nations this primary check acts chiefly by restraining marriages. The greater death-rate of infants in the poorest classes is also very important; as well as the greater mortality, from various diseases, of the inhabitants of crowded and miserable houses, at all ages. The effects of severe epidemics and wars are soon counterbalanced, and more than counterbalanced, in nations placed under favourable conditions. Emigration also comes in aid as a temporary check, but, with the extremely poor classes, not to any great extent.

There is great reason to suspect, as Malthus has remarked, that the reproductive power is actually less in barbarous, than in civilised races. We know nothing positively on this head, for with savages no census has been taken; but from the concurrent testimony of missionaries, and of others who have long resided with such people, it appears that their families are usually small, and large ones rare. This may be partly accounted for, as it is believed, by the women suckling their infants during a long time; but it is highly probable that savages, who often suffer much hardships, and who do not obtain so much nutritious food as civilised men, would be actually less prolific.

Notwithstanding that savages appear to be less prolific than civilised people, they would no doubt rapidly increase if their numbers were not by some means rigidly kept down. The Santali, or hill-tribes of India, have recently afforded a good illustration of this fact; for, as shewn by Mr. Hunter, they have increased at an extraordinary rate since vaccination has been introduced, other pestilences mitigated, and war sternly repressed. This increase, however, would not have been possible had not these rude people spread into the adjoining districts, and worked for hire. Savages almost always marry; yet there is some prudential restraint, for they do not commonly marry at the earliest possible age. The young men are often required to shew that they can support a wife; and they generally have first to earn the price with which to purchase her from her parents. With savages the difficulty of obtaining subsistence occasionally limits their number in a much more direct manner than with civilised people, for all tribes periodically suffer from severe famines. At such times savages are forced to devour much bad food, and their health can hardly fail to be injured. Many accounts have been published of their protruding stomachs and emaciated limbs after and during famines. They are then, also, compelled to wander much, and, as I was assured in Australia, their infants perish in large numbers. As famines are periodical, depending chiefly on extreme seasons, all tribes must fluctuate in number. They cannot steadily and regularly increase, as there is no artificial increase in the supply of food. Savages, when hard pressed, encroach on each other's territories, and war is the result; but they are indeed almost always at war with their neighbours. They are liable to many accidents on land and water in their search for food; and in some countries they suffer much from the larger beasts of prey. Even in India, districts have been depopulated by the ravages of tigers.

Malthus has discussed these several checks, but he does not lay stress enough on what is probably the most important of all, namely infanticide, especially of female infants and the habit of procuring abortion. These practices now prevail in many quarters of the world; and infanticide seems formerly to have prevailed, as Mr. M'Lennan has shewn on a still more extensive scale. These practices appear to have originated in savages recognising the difficulty, or rather the impossibility of supporting all the infants that are born. Licentiousness may also be added to the foregoing checks; but this does not follow from failing means of subsistence; though there is reason to believe that in some cases (as in Japan) it has been intentionally encouraged as a means of keeping down the population.

If we look back to an extremely remote epoch, before man had arrived as the dignity of manhood, he would have been guided more by instinct and less by reason than are the lowest savages at the present time. Our early semi-human progenitors would not have practised infanticide or polyandry; for the instincts of the lower animals are never so perverted as to lead them regularly to destroy their own offspring, or to be quite devoid of jealousy. There would have been no prudential restraint from marriage, and the sexes would have freely united at an early age. Hence the progenitors of man would have tended to increase rapidly; but checks of some kind, either periodical or constant, must have kept down their numbers, even more severely than with existing savages. What the precise nature of these checks were, we cannot say, any more than with most other animals.

Natural Selection. — We have now seen that man is variable in body and mind; and that the variations are induced, either directly or indirectly, by the same general causes, and obey the same general laws, as with the lower animals. Man has spread widely over the face of the earth, and must have been exposed, during his incessant migration, to the most diversified conditions. The inhabitants of Tierra del Fuego, the Cape of Good Hope, and Tasmania in the one hemisphere, and of the Arctic regions in the other, must have passed through many climates, and changed their habits many times, before they reached their present homes. The early progenitors of man must also have tended, like all other animals, to have increased beyond their means of subsistence; they must, therefore, occasionally have been exposed to a struggle for existence, and consequently to the rigid law of natural selection. Beneficial variations of all kinds will thus, either occasionally or habitually, have been preserved and injurious ones eliminated. I do not refer to strongly-marked deviations of structure, which occur only at long intervals of time, but to mere individual differences. We know, for instance, that the muscles of our hands and feet, which determine our powers of movement, are liable, like those of the lower animals, to incessant variability. If then the progenitors of man inhabiting any district, especially one undergoing some change in its conditions, were divided into two equal bodies, the one half which included all the individuals best adapted by their powers of movement for gaining subsistence, or for defending themselves, would on an average survive in greater numbers, and procreate more offspring than the other and less well endowed half.

Man in the rudest state in which he now exists is the most dominant animal that has ever appeared on this earth. He has spread more widely than any other highly organised form: and all others have yielded before him. He manifestly owes this immense superiority to his intellectual faculties, to his social habits, which lead him to aid and defend his fellows, and to his corporeal structure. The supreme importance of these characters has been proved by the final arbitrament of the battle for life. Through his powers of intellect, articulate language has been evolved; and on this his wonderful advancement has mainly depended. As Mr. Chauncey Wright remarks: "a psychological analysis of the faculty of language shews, that even the smallest proficiency in it might require more brain power than the greatest proficiency in any other direction." He has invented and is able to use various weapons, tools, traps, &c. with which he defends himself, kills or catches prey, and otherwise obtains food. He has made rafts or canoes for fishing or crossing over to neighbouring fertile islands. He has discovered the art of making fire, by which hard and stringy roots can be rendered digestible, and poisonous roots or herbs innocuous. This discovery of fire, probably the greatest ever made by man, excepting language, dates from before the dawn of history. These several inventions, by which man in the rudest state has become so pre-eminent, are the direct results of the development of his powers of observation, memory, curiosity, imagination, and reason. I cannot, therefore, understand how it is that Mr. Wallace maintains that "natural selection could only have endowed the savage with a brain a little superior to that of an ape."

Although the intellectual powers and social habits of man are of paramount importance to him, we must not underrate the importance of his bodily structure, to which subject the remainder of this chapter will be devoted; the development of the intellectual and social or moral faculties being discussed in a later chapter.

Even to hammer with precision is no easy matter, as every one who has tried to learn carpentry will admit. To throw a stone with as true an aim as a Fuegian in defending himself, or in killing birds, requires the most consummate perfection in the correlated action of the muscles of the hand, arm, and shoulder, and, further, a fine sense of touch. In throwing a stone or spear, and in many other actions, a man must stand firmly on his feet; and this again demands the perfect co-adaptation of numerous muscles. To chip a flint into the rudest tool, or to form a barbed spear or hook from a bone, demands the use of a perfect hand: for, as a most capable judge, Mr. Schoolcraft, remarks, the shaping fragments of stone into knives, lances, or arrow-heads, shews "extraordinary ability and long practice." This is to a great extent proved by the fact that primeval men practised a division of labour; each man did not manufacture his own flint tools or rude pottery, but certain individuals appear to have devoted themselves to such work, no doubt receiving in exchange the produce of the chase. Archæologists are convinced that an enormous interval of time elapsed before our ancestors thought of grinding chipped flints into smooth tools. One can hardly doubt, that a man-like animal who possessed a hand and arm sufficiently perfect to throw a stone with precision, or to form a flint into a rude tool, could, with sufficient practice, as far as mechanical skill alone is concerned, make almost anything which a civilised man can make. The structure of the hand in this respect may be compared with that of the vocal organs, which in the apes are used for uttering various signal-cries, or, as in one genus, musical cadences; but in man the closely similar vocal organs have become adapted through the inherited effects of use for the utterance of articulate language.

Turning now to the nearest allies of men, and therefore to the best representatives of our early progenitors, we find that the hands of the Quadrumana are constructed on the same general pattern as our own, but are far less perfectly adapted for diversified uses. Their hands do not serve for locomotion so well as the feet of a dog; as may be seen in such monkeys as the chimpanzee and orang, which walk on the outer margins of the palms, or on the knuckles. Their hands, however, are admirably adapted for climbing trees. Monkeys seize thin branches or ropes, with the thumb on one side and the fingers and palm on the other, in the same manner as we do. They can thus also lift rather large objects, such as the neck of a bottle, to their mouths. Baboons turn over stones, and scratch up roots with their hands. They seize nuts, insects, or other small objects with the thumb in opposition to the fingers, and no doubt they thus extract eggs and young from the nests of birds. American monkeys beat the wild oranges on the branches until the rind is cracked, and then tear it off with the fingers of the two hands. In a wild state they break open hard fruits with stones. Other monkeys open mussel-shells with the two thumbs. With their fingers they pull out thorns and burs, and hunt for each other's parasites. They roll down stones, or throw them at their enemies: nevertheless, they are clumsy in these various actions, and, as I have myself seen, are quite unable to throw a stone with precision.

As soon as some ancient member in the great series of the Primates came to be less arboreal, owing to a change in its manner of procuring subsistence, or to some change in the surrounding conditions, its habitual manner of progression would have been modified: and thus it would have been rendered more strictly quadrupedal or bipedal. Baboons frequent hilly and rocky districts, and only from necessity climb high trees; and they have acquired almost the gait of a dog. Man alone has become a biped; and we can, I think, partly see how he has come to assume his erect attitude, which forms one of his most conspicuous characters. Man could not have attained his present dominant position in the world without the use of his hands, which are so admirably adapted to act in obedience to his will. Sir C. Bell insists that "the hand supplies all instruments, and by its correspondence with the intellect gives him universal dominion." But the hands and arms could hardly have become perfect enough to have manufactured weapons, or to have hurled stones and spears with a true aim, as long as they were habitually used for locomotion and for supporting the whole weight of the body, or, as before remarked, so long as they were especially fitted for climbing trees. Such rough treatment would also have blunted the sense of touch, on which their delicate use largely depends. From these causes alone it would have been an advantage to man to become a biped; but for many actions it is indispensable that the arms and whole upper part of the body should be free; and he must for this end stand firmly on his feet. To gain this great advantage, the feet have been rendered flat; and the great toe has been peculiarly modified, though this has entailed the almost complete loss of its power of prehension. It accords with the principle of the division of physiological labour, prevailing throughout the animal kingdom, that as the hands became perfected for prehension, the feet should have become perfected for support and locomotion. With some savages, however, the foot has not altogether lost its prehensile power, as shewn by their manner of climbing trees, and of using them in other ways.

If it be an advantage to man to stand firmly on his feet and to have his hands and arms free, of which, from his pre-eminent success in the battle of life, there can be no doubt, then

I can see no reason why it should not have been advantageous to the progenitors of man to have become more and more erect or bipedal. They would thus have been better able to defend themselves with stones or clubs, to attack their prey, or otherwise to obtain food. The best built individuals would in the long run have succeeded best, and have survived in larger numbers. If the gorilla and a few allied forms had become extinct, it might have been argued, with great force and apparent truth, that an animal could not have been gradually converted from a quadruped into a biped, as all the individuals in an intermediate condition would have been miserably ill-fitted for progression. But we know (and this is well worthy of reflection) that the anthropomorphous apes are now actually in an intermediate condition; and no one doubts that they are on the whole well adapted for their conditions of life. Thus the gorilla runs with a sidelong shambling gait, but more commonly progresses by resting on its bent hands. The long-armed apes occasionally use their arms like crutches, swinging their bodies forward between them, and some kinds of Hylobates, without having been taught, can walk or run upright with tolerable quickness; yet they move awkwardly, and much less securely than man. We see, in short, in existing monkeys a manner of progression intermediate between that of a quadruped and a biped; but, as an unprejudiced judge insists, the anthropomorphous apes approach in structure more nearly to the bipedal than to the quadrupedal type.

As the progenitors of man became more and more erect, with their hands and arms more and more modified for prehension and other purposes, with their feet and legs at the same time transformed for firm support and progression, endless other changes of structure would have become necessary. The pelvis would have to be broadened, the spine peculiarly curved, and the head fixed in an altered position, all which changes have been attained by man. Prof. Schaaffhausen maintains that "the powerful mastoid processes of the human skull are the results of his erect position;" and these processes are absent in the or-ang, chimpanzee, &c., and are smaller in the gorilla than in man. Various other structures, which appear connected with man's erect position, might here have been added. It is very difficult to decide how far these correlated modifications are the result of natural selection, and how far of the inherited ef-fects of the increased use of certain parts, or of the action of one part on another. No doubt these means of change often co-operate: thus when certain muscles, and the crests of bone to which they are attached, become enlarged by habitual use, this shews that certain actions are habitually performed and must be serviceable. Hence the individuals which performed them best, would tend to survive in greater numbers.

The free use of the arms and hands, partly the cause and partly the result of man's erect position, appears to have led in an indirect manner to other modifications of structure. The early male forefathers of man were, as previously stated, probably furnished with great canine teeth; but as they gradually ac-quired the habit of using stones, clubs, or other weapons, for fighting with their enemies or rivals, they would use their jaws and teeth less and less. In this case, the jaws, together with the teeth, would become reduced in size, as we may feel almost sure from innumerable analogous cases.

In the adult male anthropomorphous apes, as Rütimeyer, and others, have insisted, it is the effect on the skull of the great development of the jaw-muscles that causes it to differ so greatly in many respects from that of man, and has given to these an-imals "a truly frightful physiognomy." Therefore, as the jaws and teeth in man's progenitors gradually become reduced in size, the adult skull would have come to resemble more and more that of existing man. As we shall hereafter see, a great reduction of the canine teeth in the males would almost cer-tainly affect the teeth of the females through inheritance.

As the various mental faculties gradually develop themselves the brain would almost certainly become larger. No one, I presume, doubts that the large proportion which the size of man's brain bears to his body, compared to the same proportion in the gorilla or orang, is closely connected with his higher mental powers. We meet with closely analogous facts with insects, for in ants the cerebral ganglia are of extraordinary dimensions, and in all the Hymenoptera these ganglia are many times larger than in the less intelligent orders, such as beetles. On the other hand, no one supposes that the intellect of any two animals or of any two men can be accurately gauged by the cubic contents of their skulls. It is certain that there may be extraordinary mental activity with an extremely small absolute mass of nervous matter: thus the wonderullly diversified instincts, mental powers, and affections of ants are notorious, yet their cerebral ganglia are not so large as the quarter of a small pin's head. Under this point of view, the brain of an ant is one of the most marvellous atoms of matter in the world, perhaps more so than the brain of a man.

The belief that there exists in man some close relation between the size of the brain and the development of the intellectual faculties is supported by the comparison of the skulls of savage and civilised races, of ancient and modern people, and by the analogy of the whole vertebrate series. Dr. J. Barnard Davis has proved, by many careful measurements, that the mean internal capacity of the skull in Europeans is 92.3 cubic inches; in Americans 87.5; in Asiatics 87.1; and in Australians only 81.9 cubic inches. Professor Broca found that the nineteenth century skulls from graves in Paris were larger than those from vaults of the twelfth century, in the proportion of 1484 to 1426; and that the increased size, as ascertained by measurements, was exclusively in the frontal part of the skull — the seat of the intellectual faculties. Prichard is persuaded that the present inhabitants of Britain have "much more capacious braincases" than the ancient inhabitants. Nevertheless, it must be admitted that some skulls of very high antiquity, such as the famous one of Neanderthal, are well developed and capacious. With respect to the lower animals, M. E. Lartet, by comparing the crania of tertiary and recent mammals belonging to the same groups, has come to the remarkable conclusion that the brain is generally larger and the convolutions are more complex in the more recent forms. On the other hand, I have shewn that the brains of domestic rabbits are considerably reduced in bulk, in comparison with those of the wild rabbit or hare; and this may be attributed to their having been closely confined during many generations, so that they have exerted their intellect, instincts, senses and voluntary movements but little.

According to a popular impression, the absence of a tail is eminently distinctive of man; but as those apes which come nearest to him are destitute of this organ, its disappearance does not relate exclusively to man. The tail often differs remarkably in length within the same genus: thus in some species of Macacus it is longer than the whole body, and is formed of twenty-four vertebræ; in others it consists of a scarcely visible stump, containing only three or four vertebræ. In some kinds of baboons there are twenty-five whilst in the mandrill there are ten very small stunted caudal vertebræ, or, according to Cuvier, sometimes only five. The tail, whether it be long or short, almost always tapers towards the end; and this, I presume, results from the atrophy of the terminal muscles, together with their arteries and nerves, through disuse, leading to the atrophy of the terminal bones. But no explanation can at present be given of the great diversity which often occurs in its length. Here, however, we are more specially concerned with the complete external disappearance of the tail. Professor Broca has recently shewn that the tail in all quadrupeds consists of two portions, generally separated abruptly from each other; the basal portion consists of vertebræ, more or less perfectly channelled and furnished with apophyses like ordinary vertebræ; whereas those of the terminal portion are not channelled, are almost smooth, and scarcely resemble true vertebræ. A tail, though not externally visible, is really present in man and the anthropomorphous apes, and is constructed on exactly the same pattern in both. In the terminal portion the vertebræ, constituting the *os coccyx*, are quite rudimentary, being much reduced in size and number. In the basal portion, the vertebræ are likewise few, are united firmly together, and are arrested in development; but they have been rendered much broader and flatter than the corresponding vertebræ in the tails of other animals: they constitute what Broca calls the accessory sacral vertebræ. These are of functional importance by supporting certain internal parts and in other ways; and their modification is directly connected with the erect or semi-erect attitude of man and the anthropomcrphous apes.

But what are we to say about the rudimentary and variable vertebræ of the terminal portion of the tail, forming the *os coccyx?* A notion which has often been, and will no doubt again be ridiculed, namely, that friction has had something to do with the disappearance of the external portion of the tail, is not so ridiculous as it at first appears. Dr. Anderson states that the extremely short tail of *Macacus brunneus* is formed of eleven vertebræ, including the imbedded basal ones. The extremity is tendinous and contains no vertebræ; this is succeeded by five rudimentary ones, so minute that together they are only one line and a half in length, and these are permanently bent to one side in the shape of a hook. The free part of the tail, only a little above an inch in length, includes only four more small vertebræ. This short tail is carried erect; but about a quarter of its total length is doubled on to itself to the left; and this terminal part, which includes the hook-like portion, serves "to fill up the interspace between the upper divergent portion of the callosites;" so that the animal sits on it, and thus renders it rough and callous. . . . Under these circumstances it is not surprising that the surface of the tail should have been roughened and rendered callous, and Dr. Murie, who carefully observed this species in the Zoological Gardens, as well as three other closely allied forms with slightly longer tails, says that when the animal sits down, the tail "is necessarily thrust to one side of the buttocks; and whether long or short its root is consequently liable to be rubbed or chafed." As we now have evidence that mutilations occasionally produce an inherited effect, it is not very improbable that in short-tailed monkeys, the projecting part of the tail, being functionally useless, should after many generations have become rudimentary and distorted, from being continually rubbed and chafed. We see the projecting part in this condition in the *Macacus brunneus,* and absolutely aborted in the *M. ecaudatus* and in several of the higher apes. Finally, then, as far as we can judge, the tail has disappeared in man and the anthropomorphous apes, owing to the terminal portion having been injured by friction during a long lapse of time; the basal and embedded portion having been reduced and modified, so as to become suitable to the erect or semi-erect position.

I have now endeavoured to shew that some of the most distinctive characters of man have in all probability been acquired, either directly, or more commonly indirectly, through natural selection. We should bear in mind that modifications in structure or constitution which do not serve to adapt an organism to its habits of life, to the food which it consumes, or passively to the surrounding conditions, cannot have been thus acquired. We must not, however, be too confident in deciding what modifications are of service in each being: we should remember how little we know about the use of many parts, or what changes in the blood or tissues may serve to fit an organism for a new climate or new kinds of food. Nor must we forget the principle of correlation, by which, as Isidore Geoffroy has shewn in the case of man, many strange deviations of structure are tied together. Independently of correlation, a change in one part often leads, through the increased or decreased use of other parts, to other changes of a quite unexpected nature. It is also well to reflect on such facts, as the wonderful growth of galls on plants caused by the poison of an insect, and on the remarkable changes of colour in the plumage of parrots when fed on certain fishes, or inoculated with the poison of toads; for we can thus see that the fluids of the system, if altered for some special purpose, might induce other changes. We should especially bear in mind that modifications acquired and continually used during past ages for some useful purpose, would probably become firmly fixed, and might be long inherited.

Thus a large yet undefined extension may safely be given to the direct and indirect results of natural selection; but I now admit, after reading the essay by Nägeli on plants, and the remarks by various authors with respect to animals, more especially those recently made by Professor Broca, that in the earlier editions of my 'Origin of Species' I perhaps attributed too much to the action of natural selection or the survival of the fittest. I have altered the fifth edition of the 'Origin' so as to confine my remarks to adaptive changes of structure; but I am convinced, from the light gained during even the last few years, that very many structures which now appear to us useless, will hereafter be proved to be useful, and will therefore come within the range of natural selection. Nevertheless, I did not formerly consider sufficiently the existence of structures, which, as far as we can at present judge, are neither beneficial nor injurious; and this I believe to be one of the greatest oversights as yet detected in my work. I may be permitted to say, as some excuse, that I had two distinct objects in view; firstly, to shew that species had not been separately created, and secondly, that natural selection had been the chief agent of change, though largely aided by the inherited effects of habit, and slightly by the direct action of the surrounding conditions. I was not, however, able to annul the influence of my former belief, then almost universal, that each species had been purposely created; and this led to my tacit assumption that every detail of structure, excepting rudiments, was of some special, though unrecognised, service. Any one with this assumption in his mind would naturally extend too far the action of natural selection, either during past or present times. Some of those who admit the principle of evolution, but reject natural selection, seem to forget, when criticising my book, that I had the above two objects in view; hence if I have erred in giving to natural selection great power, which I am very far from admitting, or in having exaggerated its power, which is in itself probable, I have at least, as I hope, done good service in aiding to overthrow the dogma of separate creations.

It is, as I can now see, probable that all organic beings, including man, possess peculiarities of structure, which neither are now, nor were formerly of any service to them, and which, therefore, are of no physiological importance. We know not what produces the numberless slight differences between the individuals of each species, for reversion only carries the problem a few steps backwards, but each peculiarity must have had its efficient cause. If these causes, whatever they may be, were to act more uniformly and energetically during a lengthened period (and against this no reason can be assigned), the result would probably be not a mere slight individual difference, but a well-marked and constant modification, though one of no physiological importance. Changed structures, which are in no way beneficial, cannot be kept uniform through natural selection, though the injurious will be thus eliminated. Uniformity of character would, however, naturally follow from the assumed uniformity of the exciting causes, and likewise from the free intercrossing of many individuals. During successive periods, the same organism might in this manner acquire successive modifications, which would be transmitted in a nearly uniform state as long as the exciting causes remained the same and there was free intercrossing. With respect to the exciting causes we can only say, as when speaking of so-called spontaneous variations, that they relate much more closely to the constitution of the varying organism, than to nature of the conditions to which it has been subjected.

Conclusion. — In this chapter we have seen that as man at the present day is liable, like every other animal, to multiform individual differences or slight variations, so no doubt were the early progenitors of man; the variations being formerly induced by the same general causes, and governed by the same general and complex laws as at present. As all animals tend to multiply beyond their means of subsistence, so it must have been with the progenitors of man; and this would inevitably lead to a struggle for existence and to natural selection. The latter process would be greatly aided by the inherited effects of the increased use of parts, and these two processes would incessantly react on each other. It appears, also, as we shall hereafter see, that various unimportant characters have been acquired by man through sexual selection. An unexplained residuum of change must be left to the assumed uniform action of those unknown agencies, which occasionally induce strongly marked and abrupt deviations of structure in our domestic productions.

Judging from the habits of savages and of the greater number of the Quadrumana, primeval men, and even their ape-like progenitors, probably lived in society. With strictly social animals, natural selection sometimes acts on the individual, through the preservation of variations which are beneficial to the community. A community which includes a large number of well-endowed individuals increases in number, and is victorious over other less favoured ones; even although each separate member gains no advantage over the others of the same community. Associated insects have thus acquired many remarkable structures, which are of little or no service to the individual, such as the pollen-collecting apparatus, or the sting of the worker-bee, or the great jaws of soldier-ants. With the higher social animals, I am not aware that any structure has been modified solely for the good of the community, though some are of secondary service to it. For instance, the horns of ruminants and the great canine teeth of baboons appear to have been acquired by the males as weapons for sexual strife, but they are used in defence of the herd or troop. In regard to certain mental powers the case is wholly different; for these faculties have been chiefly, or even exclusively, gained for the benefit of the community, and the individuals thereof have at the same time gained an advantage indirectly.

In regard to bodily size or strength, we do not know whether man is descended from some small species, like the chimpanzee, or from one as powerful as the gorilla; and, therefore, we cannot say whether man has become larger and stronger, or smaller and weaker, than his ancestors. We should, however, bear in mind that an animal possessing great size, strength, and ferocity, and which, like the gorilla, could defend itself from all enemies, would not perhaps have become social: and this would most effectually have checked the acquirement of the higher mental qualities, such as sympathy and the love of his fellows. Hence it might have been an immense advantage to man to have sprung from some comparatively weak creature.

The small strength and speed of man, his want of natural weapons, &c., are more than counterbalanced, firstly, by his intellectual powers, through which he has formed for himself weapons, tools, &c., though still remaining in a barbarous state, and, secondly, by his social qualities which lead him to give and receive aid from his fellow-men. No country in the world abounds in a greater degree with dangerous beasts than Southern Africa; no country presents more fearful physical hardships than the Arctic regions; yet one of the puniest of races, that of the Bushmen, maintains itself in Southern Africa, as do the dwarfed Esquimaux in the Arctic regions. The ancestors of man were, no doubt, inferior in intellect, and probably in social disposition, to the lowest existing savages; but it is quite conceivable that they might have existed, or even flourished, if they had advanced in intellect, whilst gradually losing their brute-like powers, such as that of climbing trees, &c. But these ancestors would not have been exposed to any special danger, even if far more helpless and defenceless than any existing savages, had they inhabited some warm continent or large island, such as Australia, New Guinea, or Borneo, which is now the home of the orang. And natural selection arising from the competition of tribe with tribe, in some such large area as one of these, together with the inherited effects of habit, would, under favourable conditions, have sufficed to raise man to his present high position in the organic scale.

[*In the next several chapters Darwin argues that the properties of man's mind are completely derivable under the terms of his evolutionary theory from behavioral features evidenced among other species. Given the challenge to this claim from even his supporters — notably Wallace, Lyell, and Gray — this case is crucial to his materialist position. Here Darwin takes pains to emphasize commonality among mammals of "lower mental powers"— e.g., fear, pleasure, and rage — and at least the precursors of "higher faculties"— e.g., curiosity, imagination, and reason. Citing evidence of their progressive appearance in children, he allows that even the highest faculties, reason and self-consciousness, could evolve through the "development and combination of the simpler ones." Eventually, he addresses the position that man is uniquely endowed by certain behavioral dispositions — viz., tool use, abstract appreciation of beauty, use of language, and belief in God — only to find even here their incipient foundations among other species. His analysis is marked by a somewhat naive anthropomorphism (e.g., finding the basis for religion in a tendency to attribute spiritual animation to nonsentient objects, and relating this to a dog's response to a windblown parasol), an approach that furnished Mivart and certain other critics with no little ammunition.*

The importance Darwin gives sexual selection in human evolution becomes clearer by the seventh chapter, where he is mainly concerned with the evolution of human races. At first, Darwin seems to find the formation of races something of an enigma. One by one he thumbs through his stock of conventional evolutionary explanations, rejecting the adequacy of each to account for the establishment of racial characters. Natural selection is disallowed because such features seem largely nonadaptive. Climate, to cite a single example, is alleged to correlate too imperfectly with skin color for it to be implicated as an agent of selection (although consideration is given the value of dark skin as protection against solar radiation). For the same reason Darwin is unprepared to advance the role of the direct effect of environment, and he finds the inherited effects of use and disuse satisfactory only "to a quite insignificant degree." Failing to discover a consistent set of adaptive principles here, Darwin at last seizes upon sexual selection to carry the day.

I have earlier offered my opinion that the outsized role Darwin gives sexual selection in Descent *represents more an infatuation than an enduring commitment to this evolutionary mechanism. Chapter VII supports my view. He moves with ungallant haste to disavow the primacy of natural selection (and its retinue of Lamarckian mechanisms) in producing racial variations; indeed, the testimony requires fewer than four pages. It is difficult to imagine Darwin in other circumstances persuaded that skin color has been indifferent to natural selection because Afrikaners over three hundred years have remained as light-skinned as their Netherlands counterparts, yet this is his own argument here. Even bearing in mind that my own judgment is undoubtedly colored by present knowledge that many of these racial features — skin color and body size and shape —are correlated in their distributions with climatic factors and are adaptive traits, the case he brings against natural selection still seems rather disingenuous. Certainly Wallace thought so.*

The middle part of Descent *is given over entirely to the subject of sexual selection. Over the course of eleven chapters Darwin elaborates lengthily the argument first set forth in* Origin: *that competition for and selection of mates have led to the establishment of sexually dimorphic traits over a wide range of animal phyla. Darwin is never quite clear whether sexual selection has generally a greater or lesser creative power than natural selection, as this passage (from Chapter VIII) indicates:*

> *Sexual selection acts in a less rigorous manner than natural selection. The latter produces its effects by the life or death at all ages of the more or less successful individuals. Death, indeed, not rarely ensues from the conflicts of rival males. But generally the less successful male merely fails to obtain a female, or obtains a retarded and less vigorous female later in the season, or, if polygamous, obtains fewer females; so that they leave fewer, less vigorous, or no offspring. In regard to structures acquired through ordinary or natural selection, there is in most cases, as long as the conditions of life remain the same, a limit to the amount of advantageous modification in relation to certain special purposes; but in regard to structures adapted to make one male victorious over another, either in fighting or in charming the female, there is no definite limit to the amount of advantageous modification; so that as long as the proper variations arise the result of sexual selection will go on. This circumstance may partly account for the frequent and extraordinary amount of variability presented by secondary sexual characters. Nevertheless, natural selection will determine that such characters shall not be acquired by the victorious males, if they would be highly injurious, either by expending too much of their vital powers, or by exposing them to any great danger. The development, however, of certain structures — of the horns, for instance, in certain stags —*

has been carried to a wonderful extreme; and in some cases to an extreme which, as far as the general conditions of life are concerned, must be slightly injurious to the male. From this fact we learn that the advantages which favoured males derive from conquering other males in battle or courtship, and thus leaving a numerous progeny, are in the long run greater than those derived from rather more perfect adaptations to their conditions of life. We shall further see, and it could never have been anticipated, that the power to charm the female has sometimes been more important than the power to conquer other males in battle.

Wallace's disagreement with Darwin over the importance of sexual selection stemmed from his concern that a more fundamental evolutionary principle — adaptation to the environment — was compromised by this explanation. Where Darwin saw a case of conspicuous male ornamentation arising from female aesthetic choice, Wallace preferred instead to see protectively cryptic female coloration arising from predator pressure. Because Wallace was even more ecologically oriented than Darwin, and because he did not share Darwin's attachment to the model of evolution from domestication (sexual selection being quite analogous to a breeding program), their differences in outlook are perhaps understandable.[4]

In the final chapters of Descent Darwin strives to apply the principles of sexual selection to man. This subject proves too refractory, however, and manages frequently to elude his grasp. We are reminded how far the naturalist has ventured into unfamiliar territory by an analysis often instantiated either by muddled and second-hand ethnographic details or not at all.

Darwin attributes almost all human secondary sex characteristics to sexual selection. These features are established by competition among males for mates or by selectivity expressed through mate choice. The greater size and strength of males receive their expected prominence as consequences of intermale competition, but Darwin overlooks an important contributory role of natural selection by his neglect (at least in this context) *of their advantages in meeting predator threat and in hunting, assuming some sexual division of labor. Differences in behavioral disposition — greater male energy, courage, pugnacity, and so on — are also held to be driven by intermale competition. Going further yet, Darwin finds males more abundantly endowed with the "higher mental faculties": observation, reason, invention, and imagination. His explanation harnesses the mechanisms of sexual selection to a Lamarckian principle that behavioral and physical traits acquired later in development tend to be transmitted in greater part to the same-sex offspring, a scheme to which he appeals in explaining why women have evolved greater beauty than men. Needless to say, these interpretations mirror a conventional Victorian world-view.*

The outcomes of sexual selection are not all markedly differentiated by sex, according to Darwin. Thus he ascribes the universal love of music and song to their primordial power in summoning the emotions that attend courtship, a courtly explanation for a phenomenon whose universality yet remains unexplained. Mating preference is claimed to have produced the worldwide variations in skin color. Although it is true that relatively slight variations in skin color within some populations have been created by sexual selection, in the main, skin-color variation is the result of natural selection on the production of melanin, a skin pigment. Melanin's proliferation in equatorial latitudes limits penetration of the damaging wave lengths of solar radiation, while its paucity in the cloudy higher latitudes admits the wave lengths wanted for vitamin D synthesis and thereby reduces the debilitating occurrence of rickets.[5] Darwin also attributes the general absence of body hair to mate preference, describing it as a "slightly injurious character"; this demonstrates his belief that sexual selection can on occasion contravene natural selection. In fact, the adaptive benefit of man's hairlessness is still poorly understood; enhanced evaporative cooling in semi-arid regions has been frequently suggested, but also challenged on empirical grounds.[6]

Finally, in considering the various cultural practices and in-

stitutions that militate against the action of sexual selection, Darwin concludes that they are widespread among extant peoples. This leaves him in an awkward position, with scant evidence for the phenomenon he is proposing (although his few illustrations of the enhanced connubial perogatives of polygamous headmen are certainly apt). In the end, he is forced to build his case chiefly upon conjectural circumstances, "At a very early period, before man attained to his present rank in the scale . . ."]

7

The Close of an Era

DESPITE its more explicitly materialistic interpretation of man's essence, *Descent* was not met with the rancor that earlier had engulfed *Origin*. In barely more than a decade the concept of evolution — even human evolution — had become installed as a familiar feature on the landscape of popular ideas. If the scientific community's judgment of the work did not always convey unbridled admiration, rarely did it concede less than sober respect. The reviews of *Descent* were for the most part favorable (Mivart's aside, of course), and the tone of criticism politely muted. A number of reviewers took the occasion to deliver the satisfying news that science posed no threat to religion after all.

In fact, a new balance was being struck between science and theology. Increasingly less wedded to a literal biblicism, established religion had begun to retract the scope of its interpretive authority over the natural universe. This drift toward liberalism cannot be laid directly to the conflict over Darwinism, which aggravated but surely did not create the internal doctrinal tensions within Christianity, particularly Anglicanism (see, for example, Himmelfarb).[1] Still, whether instrumental or not, Darwinism enjoyed full benefit of the diminished ecclesiastic claims for scriptural inerrancy. Much of evolution's widening appeal was fastened to the mistaken belief that Darwin had fashioned a paean to the Almighty (although perhaps by inadvertence). If creationism represents "design by retail," as the Reverend Henry Ward Beecher put matters, it was decidedly less impressive than producing "design by wholesale" through evolution and the natural laws it subtends. Thus, suffering the same misunderstanding, one reviewer was brought to commend *Descent* as "a system of natural Theology founded on a new basis."[2] There is some additional irony here, for this is exactly what Robert Chambers had claimed in *Vestiges of Creation* of his own much-maligned theory some twenty-five years before.

On the other side of the balance, positivism was rapidly succeeding as the regnant scientific epistemology, displacing the teleologies and providential mysteries of natural theology. While the antecedents of this shift had been established in the physical sciences for more than two centuries, Darwinism marked — and sponsored — its completion. This does not mean that all practitioners of science would thereafter abandon their previous commitments; we have already found parading under the Darwinian banner a various assemblage of opinion on these matters. However great his effort to obtain their separation, the final divorce of science from theology was not to come within Darwin's lifetime.

The *Descent* was Darwin's last major statement on evolution. Following its publication and final revised editions of *Origin* and *Variation*, Darwin retired, leaving the active advocacy of his theory in the hands of his younger supporters. During the years that followed he continued to publish on a broad variety of topics of personal interest: *The Expression of the Emotions in Man and Animals* (1872); *Insectivorous Plants* (1875); *The Effects of Cross- and Self-Fertilization in the Vegetable Kingdom* (1876); *The Different Forms of Flowers on Plants of the Same Species* (1877); *The Power of Movement in Plants* (1880); and *The Formation of Vegetable Mould, through the Action of Worms, with Observations on Their Habits* (1881). These works have been likened often to extended footnotes to his larger accomplishment; each explored some aspect of the evolutionary process, but none reshaped its earlier representations.

The retirement years passed quietly at Down, occupied by writing and the reaping of numerous honors fitting to his achievement. Although the main currents of evolutionary thought now flowed elsewhere, the influence of his labors remained. The Darwinites had gained ascendancy in the scientific societies and in the universities. Most important, the belief in organic evolution had been elevated from heresy to orthodoxy (although many, as I have noted, received it as another species of Revelation). In the spring of 1882, Darwin died. He was buried with pomp among Britain's heroes at Westminster Abbey; among his pallbearers were Hooker, Huxley, and Wallace.

If any single ambition stood unfulfilled at his death, it was the hope that natural selection be accepted as the center of the evolutionary process. Instead, the role he held out for it was met by skepticism from most quarters. Mivart's criticism on this score, for all its exaggeration on others, had not been adequately countered; even the auxiliary Lamarckian devices that Darwin had employed to augment selection under blending inheritance already were falling into discredit under the influence of Francis Galton, August Weismann (who *was* a strong supporter of selection), and others. Although many allowed an important evolutionary role to natural selection as the process producing limited adjustments of species to their environments, fewer looked to it as the exclusive mechanism by which new species originate. Indeed, the supplementation of natural selection by one or another saltatory mechanism was commonplace among the speciation theories then prevailing. It was clear that there was to be no satisfactory resolution of the evolutionary role of selection until new beach-heads were established in the area of genetics. With further advances thus stalled, upon Darwin's passing evolutionary study entered a long stage of quiescence.

8

The Rise of Modern Darwinism

AT THE BEGINNING of the twentieth century the fortunes of Darwinism had reached their lowest ebb. Surprisingly, the rediscovery in 1900 of Mendel's principles of inheritance, which much later furnished an important correction to Darwinism, initially served to discredit even further its explanation of evolution by natural selection. This curious turn of events may well be the least appreciated chapter in the brief history of evolutionary thought.

The conventional interpretation of the impact of rediscovered Mendelism holds that its ensuing union with Darwinism led directly to the flowering of modern, neo-Darwinian thinking. By this reading, the merger was accomplished through the replacement of the erroneous blending hypothesis of Darwin's original version with the new concept of the gene — an unblended, discrete unit of heredity. But the road to this marriage was far more tortuous than this account suggests.

With the emergence of classical Mendelian genetics during the first decade of this century, interest in evolution did indeed revive, but its revival wore a saltationist gown. While correctly identifying the origin of inherited variability in a process of random, spontaneous genetic mutation, the early Mendelians predicated the appearance of new species exclusively upon the rare occurrence of mutations having large-scale effects, called "macromutations" by the early geneticist Hugo De Vries. This explanation, of course, ran completely counter to Darwin's steadfast claims for speciation through small, inherited changes assembled incrementally by natural selection. With species alleged to arise by macromutation alone, the mutationists cast natural selection entirely in a negative role and reduced it to an impoverished device merely serving to eliminate the less fit. Thus a somewhat premature epitaph for Darwinism was offered by William Bateson, another of the early geneticists, who wrote in retrospect that

> The transformation of masses of population by imperceptible steps guided by selection, is, as most of us now see, so inapplicable to the facts, whether of variation or of specificity, that we can only marvel at both the want of penetration displayed by the advocates of such a proposition, and at the forensic skill by which it was made to appear acceptable even for a time.[1]

Why did not Mendelism offer grounds for reopening Darwin's case for the constructive supremacy of natural selection, perhaps along the lines that Fleeming Jenkin had unwittingly outlined earlier? Why did these early Mendelians become mutationists and not Darwinists? Only when we consider how tightly linked were then the concepts of natural selection and continuous variation does their averse reaction appear less an anomaly. Darwin had been ever-insistent that the form of variation upon which natural selection operated in its constructive role was necessarily continuous in its distribution; all forms of discrete variation —"single variations," "sports," "monstrosities," and "saltations"— he grouped indiscriminately, regardless of the relative magnitudes of their effects, and placed outside the evolutionary framework. To Darwin, evolution as a continuous process need arise from correspondingly continuous hereditary phenomena.* Given this conceptual unity of continuous var-

---

* In a perceptive psychological study, Gruber has suggested that this distinction was grounded in Darwin's associations of providential intervention with abrupt discontinuities and, contrastingly, natural process with gradual continuities.[2]

iation and the creativity of selection, it is not so strange that the mutationists found selection in this capacity incongruous with their emphasis upon discontinuous hereditary phenomena. For the same reason, neither is it coincidental that among the sturdier defenders of natural selection and Darwinism at the time were the biometricians Karl Pearson and Frank Weldon, who rejected the broad applicability of the Mendelian model of discrete genes and instead promoted a version of blending inheritance founded on their observations of the inheritance patterns of continuously varying traits — stature, cranial form, and so forth. So the supporters of Darwinism no less than its opponents found Darwinism to conflict with Mendelism. In retrospect, of course, most of the confusion here owed to a failure on both sides to distinguish between genotype and phenotype. Eventually, it would be shown that continuously varying traits could be reconciled with an underlying Mendelian structure of discrete genes, and that such phenotypes were manifestations of genes situated at multiple loci in the chromosomes. But well into the 1920s the idea persisted for many that different kinds of inheritance mechanisms were responsible for continuously and discontinuously varying traits.

It must not be understood from this too brief synopsis that the mutationist school held complete hegemony over evolutionary theory during the first decades of this century, nor that the biometricians were Darwin's only defenders. In fact, a number of non-Darwinian theories flourished, not the least being neo-Larmarckianism. Darwinism was also upheld by a number of naturalists, including August Weismann.* Still, the Darwinists notwithstanding, Mendelism took root in biology very quickly, and its principal interpreters held considerable influence on its application to evolutionary theory.

After a decade or so, mutationism as a theory of evolution

---

* The reader interested in the state and development of evolutionary thought during the first half of the century will find the volume edited by Mayr and Provine[3] an invaluable resource.

had been mostly eclipsed. Its final hurrah, Richard Goldschmidt's theory conflating the origins of species with the occasional appearance of "hopeful monsters" (his unfortunate term) through large-scale, "systemic" chromosomal rearrangement, was ill-received, among other reasons, for its extravagant denial of single-gene mutations. "The classical atomistic theory of the gene is not indispensable," he wrote, "for genetics as well as evolution. It is this theory which blocks progress in evolutionary thought. . . . We have already foreshadowed the twilight of the genes."[4] In fact, it was exactly upon small single-gene mutations, or *point-mutations*, that the reascendance of Darwinian principles was predicated.

By the 1920s the *Drosophila* experiments of Thomas Hunt Morgan, his coworkers, and students had disclosed the importance of point-mutations in regulating small-scale variation (e.g., eye color and wing form). Morgan, who earlier had endorsed the idea of speciation through great macromutational leaps, discarded this former belief but nevertheless retained (for the most part) the mutationists' disinterest with natural selection. Thus he incorrectly perceived evolution to follow mainly from the slow accumulation of recurrent point-mutations, with selection continuing in its diminished role as scourge of the unfit. However unproductive was this conception to a revised Darwinism, Morgan's work gave the modern theory its primary genetic underpinning: prolific small-scale mutations.

As it happened, most experimental geneticists of the period had too little appreciation for Darwin's population orientation to unify Darwinism with Mendelism, while the naturalists and others who did share this perspective were oblivious to the developments in genetics necessary to such a coupling. This conceptual gulf was spanned in the early 1930s, primarily in the work of R. A. Fisher, Sewall Wright, and J.B.S. Haldane, who were, naturally enough, population geneticists.* Quite in con-

---

* Provine[5] gives a full account of this formative development in evolutionary theory.

trast to their experimentalist colleagues (although Wright was also at home in the laboratory), these geneticists had adopted the biological population as their frame of reference. Collectively they attempted to explain by mathematical analysis the evolutionary changes of populations in terms of their shifting genetic constituency, as populations were impinged upon by various evolutionary factors — natural selection, mutation, gene exchange, and so on. Among their notable accomplishments, for example, they were able to show how the effects of selection, even acting at extremely low intensities, could overwhelm the contribution of recurrent mutation in the evolutionary process, thus putting to rest Morgan's claims to the contrary. In other words, they showed selection to be the major rate-governing process of evolution, and mutation's role to be the slow replenisher of the store of population variability.

The Darwinian theme of adaptive evolution is immediately recognizable in the flow of processes the population geneticists described, especially in the tendency for the genetic constituency of populations to shift in directions of ever-increasing average population fitness.* This was to prove valuable to the full union of Darwinism with Mendelism that was soon to be realized. I emphasize the word *soon* here, for none of these geneticists attempted to encompass the whole sweep of evolutionary genetics in their mathematical representations. Their interest centered mainly on the dynamics of genetic change within populations — *microevolution* — and they did not explore the process of speciation — *macroevolution* — as it might arise from the phenomena they considered. This problem — and no small one for evolutionary theory — was left for others to sort out.

With a new theoretical foundation, the reassertion of Dar-

* *Fitness* in the original Darwinian context refers to the adaptedness of organisms to their environments, although its meaning was subtly transformed by population geneticists to refer to reproductive success in the face of natural selection. Of course, better adapted organisms generally enjoy greater fertility than those more poorly adapted; this is, after all, a fundamental postulate of Darwinism.

winism gained momentum through the work of perhaps a dozen biologists who were, as Mayr aptly characterizes them, "bridge-builders" between their own fields and others — most especially genetics (for those outside).[6] One of the most outstanding examples of bridge-building was Theodosius Dobzhansky's tour de force, *Genetics and the Origin of Species*.[7] Dobzhansky was best known as a gifted laboratory geneticist, yet his interests ranged far beyond his chromosome studies to reflect the traditional emphasis in the Soviet Union (where he was trained) on natural populations. His foremost concern was to explain how processes of genetics lead eventually to new species, and he brought to this an extraordinary grasp of the interrelatedness of biological phenomena acting at different levels. Two facts in particular strike me as noteworthy about Dobzhansky's volume. First, his empirical analysis, especially that of the genetics of natural populations, is firmly planted in the conceptual framework of theoretical population genetics. Second, the large-scale features of evolution at the macroevolutionary level are extrapolated from the small-scale, microevolutionary processes. Thus the processes by which species form are tied to changes in the genetic compositions of populations, and these are traceable in turn to mathematically adduced principles of population genetics. This became a major keynote in the modern representations of Darwinism then taking form.

Others' syntheses were shortly to follow Dobzhansky's. Certainly none was so self-consciously synthetic as that of zoologist Julian Huxley (a grandson of Thomas); its title, *Evolution: The Modern Synthesis*, supplied the popular term "synthetic theory" to describe the emergent neo-Darwinian outlook.[8] That same year Ernst Mayr contributed *Systematics and the Origin of the Species*.[9] Mayr provided the neo-Darwinian synthesis its enduring definition of the species, ". . . groups of actually or potentially interbreeding natural populations, which are reproductively isolated from other such groups" (p. 120), and helped to consolidate its perspective on speciation: the process by which dispersed incipient species, initially isolated by geography, gradually acquired reproductive isolation. In this context Mayr, like Dobzhansky, emphasized the evolutionary role of whole complexes of harmoniously interacting genes (termed *coadapted* gene complexes) which, in intricately interdependent fashion, regulate the development of the entire organism. The holism of coadaptation gave the synthesis its connection between microevolutionary and macroevolutionary levels: microevolutionary processes operating differentially between geographically isolated gene pools, by rearrangement of entire integrated constellations of genes, lead eventually to the loss of systemic harmony — or mutual coadaptation — between them. The accumulative result is reproductive isolation and thus speciation. A corollary postulate is that evolutionary change ought to proceed gradually, since the need for changing large numbers of coadapted genes imparts stability to the species gene pool, which can be expected to resist admitting genetic novelties into the midst of already well-integrated gene complexes.

At least from its wholesale aspect, however, the matter of evolutionary rates was most effectively treated from the perspective of paleontology. George Gaylord Simpson was most responsible for bringing that field into the synthesis with *Tempo and Mode in Evolution*.[10] Simpson's achievement in linking paleontology to population genetics was all the more significant in light of the procrustean non-Darwinian attitudes that then encumbered paleontology. Simpson demonstrated how paleontology could be *rendered consistent* with population genetics' theoretical structure.* Considering how disparate are the universes of their respective claims, this was no minor accomplishment. Moreover, since the major, large-scale patterns of evolution — apparent long-term directionality of change, adaptive radiations (i.e., rapid proliferation of evolutionary lineages from a common ancestral stem), varied rate structures,

---

*A fine interpretation of Simpson's contribution to the modern synthesis is given by Gould.[11]

295

longevity of high taxa, and so on — are at most to be inferred indirectly if at all through observation of living forms, Simpson's contribution was necessary to a global, unified evolutionary world-view.

By 1950 all the subdisciplines of evolutionary biology were consolidated under the theoretical rubric of the modern synthesis. The synthesis served both to orient the research program of evolutionary biology and to eliminate many of the barriers and mutual misunderstandings that had grown up between its separate fields.[12] Once reached, this consensus was unbroken and relatively enduring in its form. Stephen Gould has argued — correctly, I believe — that throughout the 1950s the synthesis "hardened about its Darwinian core,"[13] meaning that the explanatory significance of random evolutionary processes diminished as selection acquired near exclusive title to the governance of evolutionary direction. A likely interpretation of this trend is that more productive research programs result from presuming the pervasiveness of selection and adaptation. Put another way, since research strategies designed to uncover nonadaptive patterns need be more exhaustive than those constructed to disclose adaptive phenomena, this circumstance leads to an "adaptationist" bias whereby nonadaptive principles are invoked mainly in the breach.

With the rise of the modern synthesis, evolutionary theory circa 1959 enjoyed an exhilarating moment of unanimity as the dominion of Darwinism was celebrated on the centennial of *Origin*. Subsequent developments, however, have eroded that consensus. It is not true that Darwinism as it became embodied by the modern synthesis now totters on the verge of collapse,[14] but neither is its posterity as the complete paradigm for organic evolution entirely secure. The challenges to the synthesis are, for the most part, concentrated on two fronts: on its conception of evolution at the genetic level and on its conception of change at the macroevolutionary level. I shall encapsulate what I consider to be some of the most important recent developments in evolutionary inquiry, significant in their own right and for their incompatibility with the neo-Darwinian synthesis, according to its critics.

The flood of advances in molecular biology during the past decade has introduced a battery of phenomena — split-genes, supergenes, multigene families, transposable chromosomal elements, etc. — with non-Mendelian properties almost completely unforeseen at the framing of the synthesis.[*] As a result, the view of the synthesis that alterations in the species gene pool (or *genome*) reflect primarily adaptive changes now seems vulnerable to attack. First, much of the evolution of DNA appears not to be clearly related to the adaptive concerns of the organism, and is perhaps indifferent to natural selection. Several recent investigators have even contended that a major component of the eukaryote genome consists of repetitive DNA regions, DNA whose existence owes only to its success in replicating and protecting itself.[15] If confirmed, this gives a measure of support to Richard Dawkins's hypothesis of the "selfish gene,"[16] a modern rephrasing of the aphorism that a chicken is the egg's way to make an egg. Viewing evolution from the gene's perspective, according to this theory genes tend to evolve that are most successful in replicating themselves, without necessary regard for the welfare of the organism in which they are packaged. While Dawkins's argument goes far beyond the phenomenon of repetitive sequences of "selfish" DNA, nevertheless such sequences — whose very existence entails an energetic burden on the organism — are notably difficult to reconcile with an adaptive view of evolution because they do not encode proteins and are therefore apparently unrepresented in the organism's phenotype.

[*] Interestingly, Watson and Crick's revolutionary publication in 1953 of the molecular basis for the genetic code had few immediate repercussions on evolutionary theory. Where classical population genetics had relied on a corpuscular model of genes arrayed on a linear chromosome (i.e., the "bead-on-a-string" model), the shift to a DNA model for the most part demanded only a conceptual miniaturization, nucleotide triplets or *codons* replacing genes as minimal units of heredity.

Next, at the level of gene products, it has become increasingly clear that natural populations maintain extraordinarily high levels of diversity in their structural proteins.[17] Such diversity was unanticipated in the synthetic view, especially since it now seems unlikely that selection alone — or even predominantly — is acting to maintain it. Many, if not most, structural proteins, for example, have regions in which the constituent amino acids vary widely within and between species; belief is growing that much of this variation is frequently unsupervised by natural selection, but instead represents the product of random mutation in concert with random sampling error in the perpetuation of the gene pool (i.e., *genetic drift*).[18] There are several reasons for this interpretation, not the least being that were selection implicated, the net intensity required to maintain this level of diversity in natural populations would be too great to sustain without threat of extinction. This is not to say, of course, that selection has no important part in the evolution of structural proteins. Even the most ardent proponents of the "neutrality theory" (so called because of the relatively limited or "neutral" evolutionary supervision attributed to selection here) recognize the functional constraints on the variability of protein structure imposed by selection. To be sure, there are proteins like hemoglobin which evidence considerable structural conservatism within and between species, and this suggests relatively few degrees freedom for nonfunctional, randomly directed evolution. On the other hand, regions of great structural diversity in more rapidly evolving proteins like albumin seem to reflect less obviously the hand of selection than more randomly fostered change. The "neutralist-selectionist" controversy grown up in recent years over this issue has mainly to do with the question of the prevalence of one mode of evolution over the other.

Evolution at the level of structural proteins, however, may have relatively little to do with that involved at the "organismal" level; King and Wilson have suggested this decoupling based on their observation that species as morphologically distinct as man and the chimpanzee have DNA sequence differences only on the order of one percent.[19] This is consistent with the broader proposal that much of the adaptive change a species undergoes — in fact, even speciation itself — may depend upon a relatively few key mutations which regulate the course of development in the organism. Some of these altered regulatory functions, however, may derive not only from the point-mutations but from chromosomal rearrangements as well.[20] Going further yet, not only is the postulate that speciation arises from the gradual accumulation of numerous small-scale mutations alleged to find little support in the molecular evidence, even the central concept of coadaptation of the gene pool has been challenged.[21] These, quite obviously, are very strong charges.

At the other end of the evolutionary spectrum, the synthesis has received several sharp assaults on its conceptions of the patterns of macroevolution. For the sake of simplicity, it is convenient to distinguish two related classes of criticism: the first *descriptive*, disputing the claim that the fossil record shows evolution rooted in continuous, adaptive change, and the second *causal*, contending that the macroevolutionary patterns are not explained by microevolutionary processes extended over a geological time-scale.

Taking up each in its turn, unquestionably no single challenge to the synthesis has provoked more attention than the theory of *punctuated equilibrium* advanced by Niles Eldredge and Stephen Gould.[22,23] They hold the regular discontinuities in the fossil record to be much less artifacts of imperfect preservation and sampling than reflections of the evolutionary process itself. Against the orthodox emphasis on long-standing adaptive trends within single lineages (*anagenesis* or *phyletic evolution*), Eldredge and Gould place the bulk of evolutionary change in discontinuous, rapid splittings of the lineages (*cladogenesis* or *speciation* in the narrow sense). In the extended intervals between such speciation events, according to this view, very little change is indicated, which implies very little adaptive change is realized outside speciation itself.

Thus, speciation and evolutionary change are compounded in a single step, while the new species resulting remain in equilibrium (or *stasis*) until either extinction eliminates them or the process repeats itself. On its face, punctuated equilibrium is an argument about the pattern of the fossil record, and as such it is descriptive. (That a more profound issue lurks beneath I shall address soon.) Opinion at present is sharply divided on its merit. Its critics tend to dwell upon the difficulties in establishing paleontological discontinuities as real and not artifactual, since this depends upon multiple negative assumptions (e.g., the geological stratigraphy *not* being discontinuous, the local fossil sequence *not* reflecting rapid regional biogeographic shifts, continuous and adaptive change *not* having occurred in fossilized soft tissues). For their part, proponents have emphasized the testability of the claim for long-term stasis, since identities of fossil forms over long gaps of time *can* be established even if the factors listed above — save the last one — intervene. It is important to remember that the hypothesis is recent and in its programmatic stage, and has not withstood the lengthy testing required to render a robust verdict on its applicability.*

---

*The only judgment I am prepared to offer is that Gould and Eldredge's use of our own species' prehistory as an exemplar was infelicitous.[23] Their claim that the morphological sequence of hominid taxa *Homo habilis/Homo erectus/Homo sapiens* is constituted of three distinctive groups of fossils which leave each taxon with pretty much the same appearance as they entered has little justification. Morphologically intermediate fossils in appropriate temporal ranges are known between each taxon, and graded change within these taxa is fairly well established. Thus, for example, the most recent *Homo erectus* specimens bear at least as close resemblance to earliest *Homo sapiens* materials as they do to the earliest fossil specimens of their own taxon. It is true that speciation has occurred within the hominids; nevertheless, only two collateral lineages can be securely indentified — our own gracile, omnivorous lineage and a now extinct robust, herbivorous one — although the robust lineage seems likely to have been further subdivided. The hominid lineage is far too fragmentary either to sustain or to reject a claim for punctuation in any of its parts, although where it is most complete, continuous — although not uniform — change seems the surer bet.

If the claim for punctuation as the dominant pattern of macroevolution is eventually sustained, what bearing has it upon the modern synthesis? It is true that punctuated equilibrium was not a prediction of the synthesis; on the contrary, Simpson emphasized continuous, phyletic evolution as the most pervasive feature of evolution at this level. At the same time, as many others have pointed out, there is an important difference between not being a *prediction* of the synthesis and not being *compatible* with it.[24] In fact, Gould has indicated that he and Eldredge devised the hypothesis in the first place to help them envision how the orthodox neo-Darwinian model for rapid speciation might appear on a geological time-scale.[25] * So punctuated equilibrium as a descriptive statement is in principle consistent with microevolutionary claims of the synthesis, but departs radically from Simpson's gradualist view of macroevolution.

Before turning to the next challenge, there is another point — assuredly more minor — that deserves to be addressed in a volume on Darwin's contribution to evolutionary theory, and this concerns the question of priority. Did not Darwin's ideas on the macroevolutionary pattern in some important way anticipate the punctuated equilibrium theory of Eldredge and Gould (see p. 304)? In fact, my own impression on reading their 1972 paper was "How clever! Somebody is mining Darwin." Gould's account of the actual origin of their inspiration is otherwise; as I have indicated, their theory is rooted in Mayr and not in Darwin. Nevertheless, there remains the question of how closely punctuated equilibrium conforms to Darwin's own understanding of the macroevolutionary pattern. Gould has re-

---

* This explanation, most closely associated with Ernst Mayr, gives a prominent role in the formation of new species to small, semi-isolated populations located at the periphery of the species range.[26] Not only are such populations bound less closely by the web of gene exchange to those of the core, in their marginal habitats they are also likely to be subject to novel and more intense selection pressures. Both circumstances, of course, ought to favor a relatively rapid evolutionary divergence of such fringe populations, which by conventional understandings leads to their eventual speciation.

cently addressed this matter as it has been raised by others:[27]

> The Darwinian claim is particularly egregious. No belief was more central to Darwin's thinking than gradualism; this may be the one point on which nearly all Darwin scholars agree (Gruber, 1974; Mayr, 1978). Darwin's faith in gradualism was greater by far than his confidence in natural selection, although he often conflated the two — as in this clearly invalid statement: 'If it could be demonstrated that any complex organ existed, which could not possibly have been formed by numerous, successive, slight modifications, my theory would absolutely break down' (1859, p. 89). To Darwin, the gradual continuity of change was part of the very definition of natural process (Gruber, 1974, pp. 125–126). One may, of course, cite the single sentence insertion into the fourth edition of the *Origin*, in which Darwin allows that lineages undergo marked fluctuation in rates, with long periods of stability and short episodes of activity, but to what effect? No one has ever believed in an absolute constancy of rates, and Darwin speaks here of anagenesis within lineages, not punctuated equilibrium. You cannot do history by selective quotation and search for qualifying footnotes. General tenor and historical impact are the proper criteria. Did his contemporaries or descendants ever read Darwin as a saltationist?[28]

Did Darwin mean what he wrote or did he not? Quite removed from the matter of Darwin's honor, which on this substantively trivial issue hardly needs my paragraph in defense (although I am certain he would wish it), is the question of the proper way to read Darwin, and here Gould and I disagree on most, but not all, major points. I agree that Darwin's texts are not to be culled selectively for passages that might supply hidden oracular meanings if vexed sufficiently; Darwin was not, after all, the Nostradamus of evolutionary biology. I agree, therefore, that Darwin need be interpreted as a whole, although not, I emphasize, as an unstratified whole. The corpus of Darwin's writings is not altogether transparent in meaning.

The later editions of *Origin*, especially, are laden with internal contradictions and departures from earlier editions — a point that at least many scholars of Darwin agree upon. In a large measure this results from his habit of emending successive editions by accretion, without bothering always to render the earlier text consistent with the alterations. In truth, as his correspondence plainly reveals, he took to the task of revision only grudgingly. Darwin's views were subject to change, like those of most scholars over their careers. For the case in point, it is true that earlier he had imagined evolution on a geological timescale to mirror its stately pace "as measured by years," despite the failure of the fossil evidence to confirm this view (see Chapter IX of *Origin*). But by the spring of 1866, when the fourth edition was prepared, Darwin finally seems to have been persuaded by Hugh Falconer's argument, drawn from analysis of fossil elephants, that "unity of type" sustained over extensive geological periods is a hallmark of the fossil record.[29] Since the persistence of form had to be reconciled with the production of variation and natural selection in order to preserve his theory, Darwin inserted these gradual processes episodically in geological time. Accordingly, we find from the fourth edition onward this new idea nestled among the older statements about the gradual tempo of evolution. Thus, I contend, we cannot rightly decide Darwin's meaning simply by an ahistorical, statistical impression of the work's "tenor"; an unweighted average of all germane passages surely provides an incorrect sense of Darwin's later thinking. Plumbing this vein further, the exact textual environment of Darwin's statements is as significant as their sequential context. We receive overwhelmingly the impression of imperceptibly gradual evolution because Darwin's characterization of evolutionary rates was bound to his preoccupation with its generation-by-generation details. After the third edition, however, whenever he did introduce a proposition about the pattern of macroevolutionary change, it was consistent with the passages in dispute; Templeton has also made this point.[30] Thus the suggestion that Darwin be understood as a saltationist confuses the issue: it was no more necessary for Darwin to have adopted saltation — speciation in a single stroke — in this context than it is for his punctuationist successors to adopt macromutations as the explanation for the macroevolutionary pattern. It is also true, as Gould states, that Darwin's statement is about anagenesis and punctuated equilibrium is about cladogenesis. In fact, Darwin's interest — or at least emphasis — centers predominantly upon anagenesis, although sometimes he seems to blur the distinction between the two (and I am unconvinced that for the present issue it is crucial). However, if Gould and Darwin both agree on long periods of stasis interrupted occasionally by rapid change (and I find no other way to interpret Darwin's statement), then a titillating if not entirely relevant possibility arises. Imagine the fossil residue sampled from a group of species evolved under Eldredge and Gould's paradigm alongside its counterpart sampled under Darwin's. Since in Darwin's revised scheme evolutionary change is no less discontinuous than in Eldredge and Gould's, the respective outcomes, I suspect, might be difficult to distinguish! Finally, that this argument of Darwin's was overlooked in the synthesis is undeniable. This fact has considerable interest for historians of science, but little bearing on the question of granting recognition for a prior formulation.*

Returning to the larger theme, a fundamentally more profound challenge to the synthesis has been offered in various recent claims that microevolutionary processes have little to do in shaping the form of macroevolution. In other words, the patterns of macroevolution are held not to result cumulatively

---

*Mendel's publications languished unread for thirty-five years before independent rediscovery of his work, while Wilhelm Weinberg's independent derivation of the Hardy-Weinberg Theorem in population genetics went virtually unrecognized for over thirty years; their priorities are undisputed. Compared to these cases, of course, Darwin's argument bears a dimmer correspondence with its well-developed successor. Perhaps the generous crediting of Goldschmidt in connection with recent theories of discontinuous speciation may furnish a more apt example.[31]

from those processes that are known to govern microevolution, and thus may have little to do with individually oriented selection. In these new formulations a strong random component is invoked to explain either the origination or extinction of species. For example, Leigh Van Valen has pointed out that lengths of survivorship for higher evolutionary taxa tend to be approximately equal, an observation apparently difficult to reconcile with any prediction from orthodox evolutionary theory.[32] It is, however, consistent with the possibility that the constituent species appear and disappear at random intervals, with little regard to their adaptive success at the microevolutionary level.* Others have shown by computer simulation that random speciation and extinction events can generate artificial macroevolutionary patterns resembling those Van Valen finds on paleontological evidence.[33] This is not to confirm, of course, that such configurations actually do arise from these random processes, but only to suggest that they might.

In a notable extension, Gould and Eldredge[13,23,25] and Stanley[34] offer a more structured explanation for higher-order evolutionary patterns, but one still placing great importance on random factors. In their view, a level of evolution operates at which species, rather than individual organisms, serve as the units of selection — hence, *species selection*. The source of variation at this level is furnished by speciation — here punctuated equilibrium is invoked — which, although it may be driven by conventional neo-Darwinian mechanisms, may equally be the result of macromutational mechanisms. What allies this evolutionary model with the ideas of Van Valen and Raup is its radical proposal for the sudden appearance of new species bearing features seemingly unrelated to (and thus undirected by) any adaptive considerations; it is this random ele-

---

* Van Valen christened this the Red Queen Hypothesis, likening the character of microevolution as the device enabling species to maintain themselves until their disappearance to Alice's breathless but always fruitless attempt to gain ground by running in *Through the Looking Glass:* the Red Queen admonishes Alice, "Now, *here*, you see, it takes all the running *you* can do, to keep in the same place."

ment that preserves the analogy between speciation and individual-level mutation. Species selection, the higher-order analogy to Darwinian selection, is credited with differentially sorting the species according to their collective properties obtained through speciation, and from this emerge the large-scale trends and patterns of the macroevolutionary level. While Gould, Eldredge, and Stanley readily concede that Darwinian selection operating at the level of individual organisms can produce some of the features upon which species selection can act, they discount the neo-Darwinist claim for evolutionary trends *exclusively* powered by the sustained influence of natural selection. To the contrary, they hold many trends to be driven by "internal" factors, unrelated at all to environmental adaptation. One example consists in the evolutionary success of sexually reproducing forms, indexed by their greater numbers of species; this "birth-bias" (meaning here a tendency to speciate prolifically) in favor of sexual species, they argue, has only to do with their enhanced facility in speciating and nothing to do with their greater adaptedness to the environment. In phenomena such as this, they contend, are found the major direction-giving forces of macroevolution.

Could it be, then, that the original radiance of the synthesis is fading? That the seminal format of the modern synthesis is inadequate to all current conceptual and empirical requirements is neither disputable nor surprising — it is no longer 1950. It has been evident for some time that a reformulation of theory in evolutionary genetics has been wanted, since — from a structural perspective — it seems that some levels of genomic evolution yield better to orthodox concepts of change than others. Such hierarchical reconstructions are presently appearing. If eventually sustained, the claims for macromutational mechanisms leading to rapid speciation — quite apart from the issue of nonadaptive speciation — would certainly require major revisions of the synthesis. On the other hand, an impressive body of evidence exists for gradual speciation on a generational time-scale,[26] along with suggestive evidence to reaffirm the concept of coadaptation.[35] But whether speciation results from few macroevolutionary changes or numerous microevolutionary changes is still far from resolution in molecular genetic terms — as I believe ultimately the question must be resolved — so that it does not appear the modern synthesis is yet disintegrating under this assault.

On the macroevolutionary front, as I have stated already, punctuated equilibrium as an empirical proposition is not, perforce, in conflict with the synthesis, although if its wide province becomes established, then a more complete theoretical explanation for stasis will certainly be wanted. Species selection, in its present form, would seem to require the most profound reworking of evolutionary theory. Of the two claims it embodies, that speciation is regularly undirected by adaptive considerations and that it furnishes grist for a higher-order sorting process, the first is unquestionably the more heretical. It is also, for reasons that I have earlier suggested, a claim that cannot be easily tested even were the world constructed along such principles as it posits, since its validation depends upon exhaustive exclusion of all adaptive possibilities. Moreover, it would seem extremely unlikely that the data of paleontology are at all well suited to such a prospect; neither are fossil species equivalent to biological species (so that speciation for a paleontologist is only loosely tied to the actual process), nor can the paleoecological context be sufficiently reconstructed to support many convincing claims about a fossil structure's lack of adaptedness. If the concept of randomly directed speciation has merit, I place my money on the assessments of those who study living populations; so far there has been no groundswell of support from their quarters. But on the separate question of higher-order sorting processes, there would seem to be no logical reason why these ought not to operate and some evidence that they might actually.[25] The concept of a hierarchy of selection levels is not altogether novel to modern evolutionary theory, but it would be disingenuous to suggest that it has been well integrated into the synthesis. It probably needs to be.

Much of the recent debate on the present and future standing of the modern synthesis has focused on the elasticity of its ideological boundaries to accommodate each new challenge. Quite expectably, its adaptability is proclaimed by its supporters and contested by its detractors. But whether the banner of the synthesis or some other flutters on the evolutionary standard is, in the end, of very little consequence to Darwin's standing. His emblem will always be affixed to that ensign. Darwin was the founder; we cannot weigh the importance of his legacy merely by the fullness with which prevailing theory reflects his original vision.

Notes

Preface

1. C. Darwin, *More Letters of Charles Darwin*, F. Darwin and A. Seward, eds. (London: Murray, 1903), vol. 2, p. 379.
2. G. de Beer, *Charles Darwin* (London: Nelson, 1963).
3. H. Gruber, *Darwin on Man*, 2nd ed. (Chicago: University of Chicago, 1981).
4. G. Himmelfarb, *Darwin and the Darwinian Revolution* (New York: Norton, 1968).
5. R. Hofstadter, *Social Darwinism in American Thought* (New York: Braziller, 1959).
6. M. Ghiselin, *The Triumph of the Darwinian Method* (Berkeley: University of California, 1969).

Autobiography

1. R. Colp, *To Be an Invalid* (Chicago: University of Chicago, 1977).

The Voyage of the Beagle

1. Himmelfarb, *Darwin*, pp. 463–464.

The Lengthy Delay

1. C. Darwin, "Essay of 1844," *Evolution by Natural Selection: Darwin and Wallace*, ed., G. de Beer (Cambridge: Cambridge University, 1958).
2. C. Darwin, *The Autobiography of Charles Darwin, 1809–1882*, ed., N. Barlow (London: Collins, 1958), p. 118.
3. R. Chambers, *Vestiges of the Natural History of Creation* (London: Churchill, 1844).
4. W. Paley, *Natural Theology* (Boston: Gould and Lincoln, 1851), p. 10.
5. Darwin, *Autobiography*, p. 122.
6. D. Hull, *Darwin and His Critics* (Cambridge: Harvard University, 1973), p. 32.
7. Darwin, *Autobiography*, p. 140.

The Origin of Species

1. D. Ospovat, *The Development of Darwin's Theory* (Cambridge: Cambridge University, 1981).
2. Himmelfarb, *Darwin*, p. 252.

The Theory Defended

1. H. Fawcett, "A popular exposition of Mr. Darwin on the Origin of Species," *Macmillan's Magazine* (1860), 3:81–92, reprinted in Hull, *Darwin and His Critics*, p. 277.
2. Himmelfarb, *Darwin*, pp. 256–261.
3. C. Lyell, *The Antiquity of Man* (London: Murray, 1863).
4. M. Ruse, *The Darwinian Revolution* (Chicago: University of Chicago, 1979), pp. 141–144.
5. Darwin and Seward, *More Letters*, vol. 2, p. 231.
6. Ghiselin, *Triumph*, 1969.
7. N. Gillespie, *Charles Darwin and the Problem of Creation* (Chicago: University of Chicago, 1979).
8. A. Sedgwick, in *The Life and Letters of Charles Darwin*, ed. F. Darwin (London: Murray, 1887), pp. 42–45.
9. T. Wollaston, "Review of *The Origin of Species*," *Annals and Magazine of Natural History* (1860), 5:132–143, reprinted in Hull, *Darwin and His Critics*, pp. 127–140.
10. R. Owen, "Darwin on the origin of species," *Edinburgh Review* (1860), 111:487–532.
11. Himmelfarb, *Darwin*, pp. 227–280.
12. Ruse, *Darwinian Revolution*, pp. 116–125, 227–228.
13. Ruse, *Darwinian Revolution*, p. 242.
14. Himmelfarb, *Darwin*, pp. 289–292.
15. F. Jenkin, "The Origin of Species," *The North British Review* (1867), 46:227–318, reprinted in Hull, *Darwin and His Critics*, pp. 303–344.
16. Darwin and Seward, *More Letters*, vol. 2, p. 379.
17. C. Darwin, *The Variation of Animals and Plants Under Domestication* (London: Murray, 1868), vol. 2, p. 243.
18. P. Vorzimmer, *Charles Darwin: The Years of Controversy* (Philadelphia: Temple University, 1970), pp. 259–261, 289 ff.
19. Vorzimmer, *Charles Darwin*, pp. 259–261, 289 ff.
20. St. G. Mivart, *The Genesis of Species* (London: Macmillan, 1871).
21. Vorzimmer, *Charles Darwin*, pp. 254–267.
22. T. Huxley, "Mr. Darwin's Critics," *Contemporary Review* (1871), 18:443–476.
23. C. Wright, "The Genesis of Species," *The North American Review* (1871), 113:63–103, reprinted in Hull, *Darwin and His Critics*, pp. 384–408.
24. Lyell, *Antiquity of Man*, 1863.
25. T. Huxley, *Evidence as to Man's Place in Nature* (London: Willams and Norgate, 1863).

The Descent of Man

1. D. Johanson and M. Edey, *Lucy: The Beginnings of Humankind* (New York: Simon and Schuster, 1981).

2. R. Leakey and R. Lewin, *Origins* (New York: Dutton, 1977).
3. S. Gould, *The Mismeasure of Man* (New York: Norton, 1981).
4. Ghiselin, *Triumph*, pp. 229–230.
5. See, for example, W. Loomis, "Skin Pigment Regulation of Vitamin D Biosynthesis in Man," *Science* (1967), 157:501–506.
6. R. Newman, "Why Man Is Such a Sweaty and Thirsty Naked Animal: A Speculative Review," *Human Biology* (1970), 42:12–27.

The Close of an Era

1. Himmelfarb, *Darwin*, pp. 392–411.
2. A. Grant, "Philosophy and Mr. Darwin," *Contemporary Review* (1871), 17:274–275, as quoted in Himmelfarb, *Darwin*, p. 358.

The Rise of Modern Darwinism

1. W. Bateson, *Problems of Genetics* (New Haven: Yale University, 1913), p. 248.
2. Gruber, *Darwin on Man*, p. 125.
3. E. Mayr and W. Provine, eds., *The Evolutionary Synthesis* (Cambridge: Harvard University, 1980).
4. R. Goldschmidt, *The Material Basis of Evolution* (New Haven: Yale University, 1940), pp. 209–210.
5. W. Provine, *The Origins of Theoretical Population Genetics* (Chicago: University of Chicago, 1971).
6. E. Mayr, "Prologue: Some Thoughts on the History of the Evolutionary Synthesis," in Mayr and Provine, *Evolutionary Synthesis*, pp. 1–48.
7. T. Dobzhansky, *Genetics and the Origin of Species* (New York: Columbia University, 1937).
8. J. Huxley, *Evolution: The Modern Synthesis* (New York: Harper, 1942).
9. E. Mayr, *Systematics and the Origin of Species* (New York: Columbia University, 1942).
10. G. Simpson, *Tempo and Mode in Evolution* (New York: Columbia University, 1944).
11. S. Gould, "G. G. Simpson, Paleontology, and the Modern Synthesis," in Mayr and Provine, *Evolutionary Synthesis*, pp. 153–172.
12. D. Shapere, "The Meaning of the Evolutionary Synthesis" in Mayr and Provine, *Evolutionary Synthesis*, pp. 388–398.
13. S. Gould, "Darwinism and the Expansion of Evolutionary Theory," *Science* (1982), 216:380–387.
14. See, for example, Gould in *Evolutionary Synthesis* or Gould in *Science*.
15. W. Doolittle and C. Sapienza, "Selfish genes, the Phenotype Paradigm and Genome Evaluation," *Nature* (1980), 284:601–603; L. Orgel and F. Crick, "Selfish DNA — the Ultimate Parasite," *Nature* (1980), 284:604–607.
16. R. Dawkins, *The Selfish Gene* (Oxford: Oxford University, 1976).
17. F. Ayala, "The Genetic Structure of Species," *Perspectives on Evolution*, R. Milkman, ed., (Sunderland, Massachusetts: Sinauer, 1982), pp. 60–82.
18. See, for example, A. Wilson et al., "Biochemical Evolution," *Ann. Rev. Biochem.* (1977), 46:573–639.
19. M.-C. King and A. Wilson, "Evolution at Two Levels in Humans and Chimpanzees," *Science* (1975) 108:107–116.
20. See, for example, G. Bush et al., "Rapid Speciation and Chromosomal Evolu-

tion in Mammals," *Proc. Natl. Acad. Sci. U.S.A.* (1977), 74:3942–3946, or G. Bush, "Stasipatic Speciation and Rapid Evolution in Animals," *Evolution and Speciation: Essays in Honor of M.J.D. White*, W. Atchley and D. Woodruff, eds. (New York: Cambridge University, 1981), pp. 201–218.

21. G. Bush, "What Do We Really Know About Speciation?" *Perspectives on Evolution*, R. Milkman, ed. (Sunderland, Massachusetts: Sinauer, 1982), pp. 119–128.

22. N. Eldredge and S. Gould, "Punctuated Equilibrium: An Alternative to Phyletic Gradualism," *Models in Paleobiology*, T. Schopf, ed. (San Francisco: Freeman 1972), pp. 82–115.

23. S. Gould and N. Eldredge, "Punctuated Equilibria; The Tempo and Mode of Evolution Reconsidered," *Paleobiology* (1977), 3:115–151.

24. G. Stebbins and F. Ayala, "Is a New Evolutionary Synthesis Necessary?" *Science* (1981), 213:967–971.

25. S. Gould, "The Meaning of Punctuated Equilibrium and Its Role in Validating a Hierarchical Approach to Macroevolution," *Perspectives on Evolution*, R. Milkman, ed., Sunderland, Massuchetts: Sinauer, 1982), pp. 83–104.

26. E. Mayr, *Animal Species and Evolution* (Cambridge: Harvard University, 1963).

27. A. Templeton and L. Giddings, "Letter to the Editor," *Science* (1981), 211:770–771; A. Templeton, "Adaptation and the Integration of Evolutional Forces," *Perspectives on Evolution*, R. Milkman, ed., (Sunderland, Massachusetts: Sinauer, 1982), pp. 15–31.

28. Gould, in Milkman *Perspectives on Evolution*, p. 84.

29. Vorzimmer, *Charles Darwin*, pp. 146–148.

30. Templeton, in Milkman, *Perspectives on Evolution*, pp. 27–28.

31. S. Gould, "Is a New and General Theory of Evolution Emerging?" *Paleobiology* (1980), 6:119–130.

32. L. Van Valen, "A New Evolutionary Law," *Evolutionary Theory* (1973), 1:1–30.

33. D. Raup, "Probablistic Models in Evolutionary Paleobiology," *American Scientist* (1977), 65:50–57.

34. S. Stanley, *Macroevolution: Pattern and Process* (San Francisco: Freeman, 1979).

35. R. Lande, "Review of S. Stanley's *Macroevolution*," *Paleobiology* (1980), 6:233–238.

Literature Cited

Ayala, F. "The Genetic Structure of Species." *Perspectives on Evolution*, ed. R. Milkman. Sunderland, Massachusetts: Sinauer, 1982.

Bateson, W. *Problems of Genetics*. New Haven: Yale University, 1913.

Bush, G. "Stasipatic Speciation and Rapid Evolution in Animals." *Evolution and Speciation: Essays in Honor of M.J.D. White*, eds. W. Atchley and D. Woodruff. New York: Cambridge University, 1981.

———. "What Do We Really Know About Speciation?" *Perspectives on Evolution*, ed. R. Milkman. Sunderland, Massachusetts: Sinauer, 1982.

Bush, G., S. Case, A. Wilson, and J. Patton. "Rapid Speciation and Chromosomal Evolution in Mammals." *Proc. Natl. Acad. Sci. U.S.A.* 74:3942–3946, 1977.

Chambers, R. *Vestiges of the Natural History of Creation*. London: Churchill, 1844.

Colp, R. *To Be an Invalid*. Chicago: University of Chicago, 1977.

Darwin, C., *The Structure and Distribution of Coral Reefs*. London: Smith, Elder, 1842.

———. *Journal of Researches into the Geology and Natural History of the Countries Visited During the Voyage of H.M.S. Beagle Round the World* (2nd ed.) London: Murray, 1845.

———. On the tendency of species to form varieties; and on the perpetuation of varieties and species by natural means of selection. *J. Proc. Linn. Soc. (Zool.)* 3:45–62, 1858.

———. *On the Origin of Species by Means of Natural Selection*. London: Murray, 1859.

———. *The Variation of Animals and Plants under Domestication* (2 vols.) London: Murray, 1868.

————. *The Expression of the Emotions in Man and Animals*. London: Murray, 1872.

————. *The Descent of Man, and Selection in Relation to Sex* (2nd ed.) London: Murray, 1874.

————. *Insectivorous Plants*. London: Murray, 1875.

————. *The Effects of Cross- and Self-Fertilization in the Vegetable Kingdom*. London: Murray, 1876.

————. *The Different Forms of Flowers on Plants of the Same Species*. London: Murray, 1877.

————. *The Power of Movement in Plants*. London: Murray, 1880.

————. *The Formation of Vegetable Mould, Through the Action of Worms, with Observations on Their Habits*. London: Murray, 1881.

————. *The Life and Letters of Charles Darwin, Including an Autobiographical Chapter* (3 vols.), ed. F. Darwin. London: Murray, 1887.

————. *More Letters of Charles Darwin* (2 vols.), eds., F. Darwin and G. Seward. London: Murray, 1903.

————. *Evolution by Natural Selection: Darwin and Wallace*, ed. G. de Beer (containing the Sketch of 1842, the Essay of 1844, and the 1888 papers of Darwin and Wallace). Cambridge: Cambridge University, 1958.

————. *The Autobiography of Charles Darwin, 1809–1882*, ed. N. Barlow. London: Collins, 1958.

————. *The Origin of Species by Means of Natural Selection. A Variorum Text*, ed. M. Peckham. Philadelphia: University of Pennsylvania, 1959.

Dawkins, R. *The Selfish Gene*. Oxford: Oxford University, 1976.

de Beer, G. *Charles Darwin*. London: Nelson, 1963.

Dobzhansky, T. *Genetics and the Origin of Species*. New York: Columbia University, 1937.

Doolittle, W. and C. Sapienza. "Selfish Genes, the Phenotype Paradigm and Genome Evolution." *Nature* 284:601–603, 1980.

Eldredge, N. and S. Gould. "Punctuated Equilibrium: An Alternative to Phyletic Gradualism," *Models in Paleobiology*, ed. T. Schopf, San Francisco: Freeman, 1972.

Fawcett, H. "A Popular Exposition of Mr. Darwin on the Origin of Species." *Macmillan's Magazine* 3:81–92, 1860.

Gillespie, N. *Charles Darwin and the Problem of Creation*. Chicago: University of Chicago, 1979.

Ghiselin, M. *The Triumph of the Darwinian Method*. Berkeley: University of California, 1969.

Goldschmidt, R. *The Material Basis of Evolution*. New Haven: Yale University, 1940.

Gould, S. "G. G. Simpson, Paleontology, and the Modern Synthesis," *The*

Evolutionary Synthesis, eds., E. Mayr and W. Provine. Cambridge: Harvard University, 1980.

————. "Is a New and General Theory of Evolution Emerging?" *Paleobiology* 6:119–130, 1980.

————. *The Mismeasure of Man*. New York: Norton, 1981.

————. "Darwinism and the Expansion of Evolutionary Theory." *Science* 216:380–387, 1982.

————. "The Meaning of Punctuated Equilibrium and Its Role in Validating a Hierarchical Approach to Macroevolution." *Perspectives on Evolution*, ed. R. Milkman. Sunderland, Massachusetts: Sinauer, 1982.

Gould, S. and N. Eldredge. "Punctuated Equilibria: The Tempo and Mode of Evolution Reconsidered." *Paleobiology* 3:115–151, 1977.

Grant, A. "Philosophy and Mr. Darwin." *Contemporary Review* 17:274–275, 1871.

Gruber, H. *Darwin on Man* (2nd ed.) Chicago: University of Chicago, 1981.

Himmelfarb, G. *Darwin and the Darwinian Revolution*. New York: Norton, 1968.

Hofstadter, R. *Social Darwinism in American Thought*. New York: Braziller, 1959.

Hull, D. *Darwin and His Critics*. Cambridge: Harvard University, 1973.

Huxley, J. *Evolution: The Modern Synthesis*. New York: Harper, 1942.

Huxley, T. *Evidence as to Man's Place in Nature*. London: Williams and Norgate, 1863.

————. "Mr. Darwin's Critics." *Contemporary Review* 18:443–476, 1871.

Jenkin, F. "The Origin of Species." *The North British Review* 46:227–318, 1867.

Johanson, D. and M. Edey. *Lucy: The Beginnings of Humankind*. New York: Simon and Schuster, 1981.

King, M.-C. and A. Wilson. "Evolution at Two Levels in Humans and Chimpanzees." *Science* 108:107–116, 1975.

Lande, R. Review of S. Stanley's *Macroevolution*. *Paleobiology* 6:233–238, 1980.

Leakey, R. and R. Lewin. *Origins*. New York: Dutton, 1977.

Loomis, W. "Skin Pigment Regulation of Vitamin D Biosynthesis in Man" *Science* 157:501–506, 1967.

Lyell, C. *The Principles of Geology* (3 vols.) London: Murray, 1830–1833.

————. *The Antiquity of Man*. London: Murray, 1863.

Mayr, E. *Systematics and the Origin of Species*. New York: Columbia University, 1942.

————. *Animal Species and Evolution*. Cambridge: Harvard University, 1963.

————. Prologue: Some Thoughts on the History of the Evolutionary Synthesis, *The Evolutionary Synthesis*, eds., E. Mayr and W. Provine. Cambridge: Harvard University, 1980.

Mayr, E. and W. Provine, eds. *The Evolutionary Synthesis*. Cambridge: Harvard University, 1980.

Mivart, St. G. *The Genesis of Species*. London: Macmillan, 1871.

Newman, R. "Why Man Is Such a Sweaty and Thirsty Naked Animal: A Speculative Review." *Human Biology* 42:12–27, 1970.

Orgel, L. and F. Crick. "Selfish DNA—the Ultimate Parasite." *Nature* 284:604–607, 1980.

Ospovat, D. *The Development of Darwin's Theory*. Cambridge: Cambridge University, 1981.

Owen, R. Darwin on the origin of species. *Edinburgh Review* 111:487–532, 1860.

Paley, W. *Natural Theology*. Boston: Gould and Lincoln, 1851.

Provine, W. *The Origins of Theoretical Population Genetics*. Chicago: University of Chicago, 1971.

Raup, D. "Probablistic Models in Evolutionary Paleobiology." *American Scientist* 65:50–57, 1977.

Ruse, M. *The Darwinian Revolution*. Chicago: University of Chicago, 1979.

Sedgwick, A. In *The Life and Letters of Charles Darwin*, ed. F. Darwin. London: Murray, 1887.

Shapere, D. "The Meaning of the Evolutionary Synthesis." *The Evolutionary Synthesis*, eds., E. Mayr and W. Provine. Cambridge: Harvard University, 1980.

Simpson, G. *Tempo and Mode in Evolution*. New York: Columbia University, 1944.

Stanley, S. *Macroevolution: Pattern and Process*. San Francisco: Freeman, 1979.

Stebbins, G. and F. Ayala. "Is a New Evolutionary Synthesis Necessary?" *Science* 213:967–971, 1981.

Templeton, A. "Adaptation and the Integration of Evolutional Forces." *Perspectives on Evolution*, ed. R. Milkman. Sunderland, Massachusetts: Sinauer, 1982.

Templeton, A. and L. Giddings. "Letter to the Editor." *Science* 211:770–771, 1981.

Van Valen, L. "A New Evolutionary Law." *Evolutionary Theory* 1:1–30, 1973.

Vorzimmer, P. *Charles Darwin: The Years of Controversy*. Philadelphia: Temple University, 1970.

Wallace, A. On the tendency of varieties to depart indefinitely from the original type. *J. Proc. Linn. Soc. (Zool.)* 3:53–62, 1858.

Wilson, A., S. Carlson, and T. White. "Biochemical Evolution." *Ann. Rev. Biochem.* 46:573–639, 1977.

Wollaston, T. Review of *The Origin of Species. Annals and Magazine of Natural History* 5:132–143, 1860.

Wright, C. The Genesis of Species. *The North American Review* 113:63–103, 1871.

Index

Page numbers in roman refer to Darwin's works; page numbers in italics refer to Dr. Jastrow's or Professor Korey's discussion.